Autorenkollektiv

Inorganic Chemistry

The Problems for the Two-electron Bond in Inorganic Compounds; Analysis of the Coordination Number *N*

Ch. K. Jørgensen

Cationic and Anionic Complexes of the Noble Gases

H. Selig and J. H. Holloway

Extraction of Metals from Sea Water

K. Schwochau

THEORIA CUM PRAXI

AKADEMIE-VERLAG · BERLIN

Autorenkollektiv

Inorganic Chemistry

Autorenkollektiv

Inorganic Chemistry

With 24 Figures and 35 Tables

AKADEMIE-VERLAG · BERLIN
1984

Die Originalausgabe erscheint im Springer-Verlag Berlin · Heidelberg · New York · Tokyo
in der Schriftenreihe „Topics in Current Chemistry", Volume 124

Vertrieb ausschließlich für die DDR und die sozialistischen Länder

Erschienen im Akademie-Verlag, DDR-1086 Berlin, Leipziger Straße 3—4
Lizenznummer: 202 · 100/534/84
Printed in the German Democratic Republic
Gesamtherstellung: VEB Druckerei „Thomas Müntzer", 5820 Bad Langensalza
LSV 1255
Bestellnummer: 763 439 3 (6880)
06800

Table of Contents

The Problems for the Two-electron Bond in Inorganic Compounds; Analysis of the Coordination Number *N*

Christian Klixbüll Jørgensen

Département de Chimie minérale, analytique et appliquée, University of Geneva, CH 1211 Geneva 4, Switzerland

Table of Contents

1 Electricity, Electrons and Chemistry

Lewis [1] wrote in 1916 a paper suggesting that chemical bonds are effectuated by electron-pairs, and that well-behaved elements have 8 outer (valency) electrons distributed in four pairs, either shared with adjacent atoms to form chemical bonds, or remaining on the atom as "lone-pairs". With exception of helium, the noble gases have also 8 electrons, but they are on the limit of becoming inner electrons (like in the subsequent alkaline-metals) and lack typical characteristics of lone-pairs, such as proton affinity in solution (however much EH^+ and EX^+ formed by the halogens can be detected in mass-spectra).

One of the major difficulties for the Lewis paradigm is the Dalton opinion that "all compounds consist of molecules". This is generally valid in the gaseous state at reasonable temperature and pressure (say, below 1000 °C and 20 atm.) in the absence of violent electric discharges. It is also true for the typical organic compounds (though many biological and synthetic polymers may have to be excluded) in condensed matter (liquids, and vitreous, amorphous and crystalline solids) though the Dalton opinion (via the modification introduced by Arrhenius) must be extended to include polyatomic cations and anions such as $N(CH_3)_4^+$ and $B(C_6H_5)_4^-$ not formed by simple addition or dissociation of protons in neutral molecules.

In view of several million organic compounds being characterized, it cannot be disputed that the Dalton opinion is valid for the majority of all compounds; but it is hardly the majority of all inorganic compounds. For reasons to become evident below, it is very difficult to count the inorganic compounds in Gmelin the same way as it is done in Beilstein, and their number may not be defined within a factor 2. The same difficulty occurs to a stronger extent in minerals. The Lewis description works marvellously for alkanes, tetra(alkyl)ammonium ions, diamond and adamantoid crystals such as gallium arsenide and indium antimonide not containing distinct molecules (though it is sometimes argued that the whole crystal is a molecule). Already at the end of last Century, the organic chemists developed a coherent description of double bonds (in olefins and carbonyl groups) and triple bonds (in acetylene and nitriles) which was incorporated in the Lewis paradigm. The writer has no objection to speak about triple bonds in the isoelectronic species N_2, CO and CN^- but is very reluctant to speak about multiple bonds in nearly all inorganic species. It is very difficult to find experimentally differing consequences for OPX_3 between a double bond between oxygen and phosphorus and a single (so-called semi-polar) bond between O^- and P^+. Any how, the main problems for the Lewis paradigm applied to inorganic chemistry occur, when 5 or more neighbor atoms is bound to a given atom, such as in the octahedral species PF_6^-, SF_6, ClF_6^+, $Co(NH_3)_6^{+3}$, $Sb(OH)_6^-$, XeO_6^{-4}, WF_6, $W(CH_3)_6$ and $PtCl_6^{-2}$. Since the 6 chemical bonds in these examples are persistent and obviously equivalent, it is usual to speak about "deviations from the octet rule". However, the more important part of the Lewis paradigm, two electrons per bond, is maintained, and people look for the 12 electrons participating in the bonding. As discussed in this review, as well as in previous papers [2-5] this line of thought terminates in rather ludicrous conclusions.

It may be useful to look into the way electrons became relevant to chemistry. In the century between Boyle and Lavoisier, the concept "chemical element" was consolidated, and in the system [6] elaborated by Lavoisier the last few years before his execution, matter has an absolutely invariant mass distributed on non-transmutable elements each characterized by their atomic weight (probably an infinite decimal fraction). Lavoisier assumed about 30 elements to exist, among which some materials correctly considered [7] as oxides of metals to be isolated subsequently (such as the alkaline earths magnesia, lime, strontia and baryta). He did not make the same proposition for sodium and potassium, because he suspected all bases to contain nitrogen (like ammonia and the amines). By the same token, he assumed all acids to contain oxygen, providing the name muriatic acid for hydrochloric acid (after all, murium might have been fluorine, as far not too precise atomic weights go). Seen in hindsight, it would have been better to call the lightest element oxygen and $Z = 8$ hydrogen (representing 88% of the weight of water).

Anyhow, it is important for our purpose that Lavoisier also recognized principles without any weight. The most widely distributed principle is heat (in a calorimetric context, without transformation of kinetic energy) and another light. There was a long-lasting controversy as to whether there is one electricity (occurring in deficit or excess relative to neutral bodies) or two (the positive and negative charges interacting electrostatically according to Coulomb in a way analogous to imaginary masses in Newton's gravitation). Like the energy added to the rest-mass by Einstein in 1905, the balances were not at all able to detect any electric mass, and Lavoisier was certainly justified from an experimental point of view to consider that chemical reactions do not modify the total mass, and to classify the electricities as a principle.

Until Schrödinger proposed his equation in 1925, chemistry did not present any possibility of becoming incorporated as a sub-division of physics. The well-established facts known about chemical bonding were distinctly incompatible with Newton's mechanics and with the electromagnetic theory of Maxwell. Nevertheless, the electrolysis of molten anhydrous salts (started 1807 by Davy preparing alkaline and alkaline-earth metals) inspired Berzelius to consider halides as bound to metallic elements by an attraction between negative and positive charges, and this picture was rendered quantitative by Faraday showing the transfer of a small integer times a mole of electrons (96485 coulomb) when a mole of a substance undergoes electrochemical reactions. This is the origin of the habit [8] among inorganic chemists to describe oxidation states by positive or negative integers, whereas organic chemists following Kekulé and Couper tend to talk about valency [9] as a positive number. There are two major components of this discrepancy, one being easy to remedy by considering hydrocarbons as containing hydrogen in the oxidation state H(—I) like halides (arguments can be given [8] that H(I) only occurs bound to nitrogen, the chalkogens and the halogens, but H(—I) in non-metallic compounds of all other elements). The other is a more profound difficulty due to the catenation of carbon (and to a lesser extent by sulfur and other elements) that ethane in no sense contains C(III) like monomeric $C(C_6H_5)_3$ studied by Gomberg, nor ethylene C(II) like CO.

The system of Lavoisier of weightless principles and 76 distinct elements [7] still was flawless in 1896 when radioactivity was discovered [10,11]. Since, then our ideas of the constitution of matter (with positive rest-mass) have undergone two fundamental reconstructions [12-14], one in the period 1911–1932 where nuclei are characterized by the two positive integers Z and A, since the positive charge (determining the chemical behavior) is Z times the protonic charge e, and A is close to the atomic weight falling (on the scale of 12 for the abundant carbon isotope) between (A — 0.10) and $(A + 0.05)$ according to well-known regularities [15]. Since 1932 until recently, an absolute sense was ascribed to Z protons and $(A - Z)$ neutrons being the constituents of nuclei in spite of the total rest-mass typically being 0.8% decreased. Since 1964, a second reconstruction describes "elementary particles" as either being leptons (such as electrons and muons) or consisting of quarks [12].

In 1897, Thomson proved that the electrons liberated by applying a high voltage to the cathode of a Crookes tube are identical (and do not depend on the cathode element) and have a tiny atomic weight 0.00055 (though it was soon realized that this value increases with kinetic energy and is twice as large, if the electron has passed a potential difference of 511000 volt). These facts about electrons constituted the first bridge between principles and elements. One of the two electricities is not entirely without mass, and is the material agent of redox reactions, much like phlogiston was assumed to be before Lavoisier. As a further example that chemical concepts never fade away completely (and that there often is a part of the truth in both sides of a heated controversy) it may be added that Lavoisier obviously was right in considering oxides as compounds, but also that the electric conductivity of alkaline and coinage metals is enacted by phlogiston surrounding small cations, and that the dark blue, dilute solutions of reactive metals in liquid ammonia in a way contain dissolved phlogiston.

Since around 1930, it is perfectly clear that chemistry is determined by electronic densities surrounding nuclei (characterized by their charge Z) which (for chemical purposes) can be approximated as geometrical points, though certain experimental techniques, such as hyperfine structure of electron spin resonance, as well as nuclear magnetic and quadrupolar resonances, go beyond this approximation. The main reason why electrons play this decisive role in chemistry is their exceptionally low atomic weight, corresponding to a very large quantum-mechanical kinetic energy (when the electrons are confined in a small volume) which is inversely proportional to the square of the linear dimension of the confining volume [16] under equal circumstances. One of the corollaries of this kinetic energy is that it prevents implosion of both molecules and ionic salts and contributes significantly, to determine the optimized manifold of internuclear distances.

2 Oxidation States and the Periodic Table

Quantitative analysis only determines the equivalent weight (e.g. the amount of a given element combined with 35.45 g chlorine in a chloride) which has to be multiplied by z (what we now [8] call the oxidation number) in order to obtain the atomic weight. Since Z was first evaluated from X-ray spectra 1913, the only distinctive parameter for an element was its atomic weight, and only when a general consensus was obtained in favor of Avogadro's hypothesis at the chemical congress in Karlsruhe 1860, it was possible to elaborate the Periodic Table. The most thorough version was proposed by Mendeleev 1869, but six or eight contemporary chemists [17] suggested related versions.

Many details of this masterpiece of chemical induction [10, 18] were decided by pushing from firmer evidence for adjacent elements. Thus, beryllium had to be accommodated in column II, because boron needed a space in III; the equivalent weights

59.5 and 39.7 of uranium in its two oxidation states were interpreted as $A = 119$ and U(II) and U(III) (not entirely unreasonable midway between aluminum and thallium) until indium was discovered, needing a further space in III and propulsing U(IV) and U(VI) up to $A = 238$ at the last position, following $A = 232$ for thorium. As a definition, the highest (but not necessarily the most frequent) oxidation state should provide the column number. Thus, Mn(VII), Pb(IV) and Bi(V) create no problems, though they are not particularly representative for their elements. Mendeleev could not avoid some exceptions to this definition. He felt that absence of high z for oxygen and fluorine was not a serious argument, because no elements have even higher electronegativities (as we would say). The non-existence of bromine(VII) was much discussed [19] until BrO_4^- was prepared in 1968 and turned out to be nearly as stable as perchlorate. However, copper(II) and gold(III) were familiar to Mendeleev, who, originally constructed three tetrade groups (Fe, Co, Ni, Cu; Ru, Rh, Pd, Ag; Os, Ir, Pt, Au) because of this. In spite of the entirely different chemical behavior of the alkaline and the coinage metals, he finally yielded to a pressure for 7 columns and three triades. When argon was discovered 1894, he first tried to explain it as N_3 (in analogy to ozone O_3) but when (the much lower concentrations) of neon, krypton and xenon were isolated from atmospheric air, he installed a column VIII for these noble gases and the triades. Only three of these elements are known with $z = 8$, oxides (and not fluorides) of ruthenium, osmium and xenon.

Following the ideas of Berzelius, Abegg proposed a general law that an element at most varies its oxidation state by 8 units, such as N(—III) to N(V), S(—II) to S(VI) and Cl(—I) to Cl(VII), imitated by the higher Z values in the same column. The physical reasoning behind this constant 8 remained enigmatic until Stoner in 1924 rationalized the Periodic Table as consecutive filling (Aufbauprinzip) of nl-shells each able to accommodate at most $(4l + 2)$ electrons. There is no doubt today [8,16] that the "octet rule" of Lewis is connected with the sum $2 + 6 = 8$ of ns and np shells, whereas one would expect 10 for the d shell. Actually, four such exceptions from Abegg's rule are now known for manganese and rhenium changing from M(—III) in (pyrophoric) $M(CO)_4^{-3}$ to M(VII) in MO_4^- and ruthenium and osmium changing from M(—II) in $M(PF_3)_4^{-2}$ to M(VIII) in MO_4. One should not put too much emphasis on both sulfur and chromium varying from M(—II) to M(VI). Whereas sulfate and chromate show a quite similar chemistry, Cr(—II) is not a monatomic anion like sulfide, but occurs in complexes of CO and PF_3 like other negative oxidation states of d-group elements. The only well-characterized monatomic d-group anion is gold(—I) (in semi-conducting CsAu) isoelectronic with the mercury atom, thallium(I), lead(II) and bismuth(III).

In 1916 (the same year as Lewis published his paper on electron-pairs) Kossel [20,21] proposed a generalized treatment of inorganic chemistry based on relations with X-ray and atomic spectra. As one would expect, this study emphasized the electrovalent aspects of bonding and introduced an effective number of electrons $K = (Z - z)$ as the difference between the atomic number Z and the oxidation state z. This expression would be exact for a gaseous monatomic ion with charge z, and Kossel argued that it was a good approximation in many compounds. It turns out that not only the isoelectronic series with $K = 2$, 10, 18, 36, 54 and 86 (like the noble gases) contain above ten differing oxidation states known [8,21] in non-catenated compounds, such as C(—IV) to Cl(VII) for $K = 10$ or Ge(—IV) to Ru (VIII) for $K = 36$, but also other *K*-values (corresponding to closed d- or f-shells) provide very long isoelectronic series, such as Mn(—III) to Br(VII) for $K = 28$; Ru(—II) to Xe(VIII) for $K = 46$; and Yb(II) to Os(VIII) for $K = 68$. Most other *K*-values have only three to five members in their isoelectronic series, though $K = 74$ stretches from Ta(—I) to Au(V) in AuF_6^-.

It is well-known from "ligand field" theory [8,16] that $(K - 18)$ d-like electrons occur in iron-group compounds, as well as $(K - 36)$ in the 4d and $(K - 68)$ in the 5d group compounds. The six d-like electrons in octahedral complexes with $K = 74$ are almost non-bonding, though they may be stabilized by back-bonding in $Ta(CO)_6^-$ and $W(CO)_6$. These d^6 systems have the total spin-quantum number *S* zero in the groundstate, corresponding to diamagnetic behavior. Other combinations of *K* in definite local symmetries may be particularly stabilized with positive *S* (by itself a deviation from the Lewis paradigm) such as d^3 with $S = 3/2$ in octahedral complexes, where the three approximately non-bonding d-like orbitals each have one electron. The large majority of the manganese(II) and iron(III) compounds with $K = 23$ have $S = 5/2$.

The high *S* values observed in many 3d group compounds constitute an intermediate situation between main-group compounds and monatomic ions [8,21-24]. In the 4f group mainly represented by Ce(III) to Yb(III) the similarity with atomic spectra is striking [25-29] and $4f^7$ represented by Eu(II), Gd(III) and Tb(IV) always showing $S = 7/2$ have the highest known *S* value in compounds (which can even be determined in metallic alloys of europium and gadolinium and in the elements). The situation is slightly less clear-cut in the 5f group showing some characteristics [18,30,31] intermediate between the 3d and the 4f groups.

As far goes the variation of the oxidation state z for a given d-group element, there is a certain discontinuity between manganese and iron, and at one element later in the 4d and 5d groups (between ruthenium and rhodium, and between osmium and iridium, however much this is inside the triades). Before the discontinuity [32,33] the z values readily vary, and can be high, up to the maximum value ($K = 18$, 36 or 68). After the discontinuity, z varies much less, and tends to keep at moderate values such as 2 and 3 in the 3d group, and 3 or 4 in the two next groups. The 4f group has a very pronounced propensity [28,34] for M(III) but the 5f group shows a discontinuity [10,33] between neptunium and plutonium rather similar to the break between manganese and iron. M(III) is stabilized from americium to mendelevium, but nobelium ($Z = 102$) is almost impossible to oxidize beyond M(II) in analogy [32] to silver(I) being much more difficult to oxidize than copper(I).

3 Coordination Number *N* from Werner to Bjerrum

It can still be seen in mineralogical museums how Berzelius classified many compounds as adducts of oxides, as when alum is written $K_2O \cdot Al_2O_3 \cdot 4\,SO_3 \cdot 24\,H_2O$ whereas it was later written $KAl(SO_4)_2 \cdot 12\,H_2O$ and the crystal structure indicates only half of the water molecule oxygen atoms directly connected to aluminum $K[Al(OH_2)_6](SO_4)_2(OH_2)_6$. Actually, the large majority of minerals are mixed oxides (and though these formulae derive from the precursor ideas of electrovalent bonding, considering calcium sulfate as an adduct $CaO \cdot SO_3$, they have the undoubted advantage to be directly compared with results of quantitative analysis) and a great part of the others are mixed sulfides.

In the laboratory, a great number of "double salts" were prepared, among which the alums (with a great choice of univalent and trivalent metallic elements) are a good example. According to Arrhenius, their solutions dissociate completely to ions such as K^+ and hydrated M^{+3}. Also "salt hydrates" can show a rapidly changing number of moles of water, though certain dehydrations of inorganic compounds, as $K^+HOSO_3^-$ forming pyrosulfate $O_3SOSO_3^{-2}$ or $(K^+)_2HOPO_3^{-2}$ forming pyrophosphate $O_3POPO_3^{-4}$ are rather irreversible, and quite comparable to the organic dehydrations (performed with concentrated sulfuric acid) of formic acid to CO or of ethanol to di(ethyl)ether and eventually to ethylene. However, some "double salts" show similar fairly irreversible changes, and 4 KCN · $Fe(CN)_2$ hardly show any of the expected reactions of CN^- and hydrated Fe^{+2} though the aqueous solution alright contains K^+.

Alfred Werner started chemical investigations with stereochemistry of organic nitrogen compounds and got skeptical against the chains of NH_3 molecules in "robust" (slowly reacting) complexes proposed by Blomstrand in analogy to CH_2 chains in *n*-alkanes (before Arrhenius, ammonium salts XNH_4 were thought to have five atoms directly bound to nitrogen). This chain model was defended by S. M. Jørgensen carefully characterizing a great number of chromium(III), cobalt(III) and rhodium(III) complexes [35]. Long time before any crystal structures were determined, Werner [36,37] proposed in 1893 that these cations, molecules or anions systematically had six ligating atoms bound to the central atom in an octahedron playing the same rôle as the regular tetrahedron in carbon chemistry. Other central atoms, such as palladium(II), platinum(II) and gold(III) have four ligating atoms bound in a square, though isolation of robust *trans*- and *cis*-isomers MX_2Y_2 is mainly successful for Pt(II) only.

There is no reason to discuss here what Werner [36] really meant with Hauptvalenz and Nebenvalenz, and it is likely that he introduced the words as an etape of a psychological evolution, rapidly arriving at the conclusion that there is no experimentally detectable difference between the six chloride ligands in $PtCl_6^{-2}$ (as verified 1922 by the crystal structure of cubic K_2PtCl_6). Already in main-group chemistry, sulfates and perxenates are much more familiar than SO_3 and XeO_4, and by the same token, $PtCl_6^{-2}$ and $IrCl_6^{-2}$ are much better known than $PtCl_4$, and $IrCl_4$ does not exist according to Delépine. Conceptually, there was at the beginning of this century a belief that binary compounds are simpler than ternary and quaternary compounds such as $K_3Co(NO_2)_6$. The experimental work of Werner showed that this is not always the case, and at least implicitly, the idea of an inorganic chromophore [38] such as $Co(III)N_6$ in complexes of unidentate, bidentate (among which the most popular is ethylenediamine $NH_2CH_2CH_2NH_2$ having the more systematic name 1,2-diamino-ethane) and multidentate ligands; or $Rh(III)Cl_6$ occurring both in salts of $RhCl_6^{-3}$ and in solid $RhCl_3$ having properties comparable to anhydrous $CrCl_3$ is a better scheme of classification for many purposes than the mere question of binary or ternary compounds. Already Werner studied complexes of biological, deprotonated amino-acids such as glycinate, and a large number of synthetic multidentate amino-polycarboxylates have since been developed, a prime case being the potentially sexidentate ligand ethylenediaminetetra-acetate of which Schwarzenbach studied complexes of nearly all the elements, and which allow the absorption spectra to be measured [39] of several $Co(III)N_2O_4$ and $Ni(II)N_2O_4$ chromophores. Even when the formation of such complexes is not particularly exothermic, the change of free energy is very favorable, because of the "chelate effect" on the entropy.

Ammonia is one of the most interesting ligands to study in aqueous solution, and the conceptually simplest technique is to measure the ammonia vapor pressure (at a strictly defined temperature) indicating the molar concentration $[NH_3]$ of free ligand. The difference between the amount of NH_3 added to the solution of the complex-forming ion, and $[NH_3]$ can then be interpreted as the average number $\bar{n}$ of ligands bound, multiplied by the concentration of the central atom. However, Jannik Bjerrum[40,41] applied pH measurements (with a glass electrode) combined with the mass-action result

$$pH = 9.4 - \log_{10}[NH_4^+] + \log_{10}[NH_3] \quad (1)$$

rendered valid (on a concentration basis) by the the presence of a large concentration of a non-complexing (say, ammonium or sodium) salt. Bjerrum emphasized that $\bar{n}$ as a function of $[NH_3]$ is independent of the central-atom concentration, if only monomeric complexes are formed (oligomerization is the rule rather than the exception for hydroxo complexes) and that such a "formation curve" allows the detection of all the five mixed complexes $Ni(NH_3)_n(OH_2)_{6-n}^{+2}$ which had not all been isolated as salts. The two geometrical isomers for n = 2, 3 and 4 cannot be distinguished that way. Analogous results can be obtained for ethylenediamine, where one has to take into account the two pK values (close to 7.5 and 10) for the protonated forms of the ligand. Palladium(II) represents at two points a limiting situation[42] both because one has to obtain equilibrium from both sides in order to obtain significant formation constants, and because one cannot determine $\log_{10}K$ much higher than the pK value 9.4 in Eq. (1). Actually, the four $\log_{10}K_n$ = 9.6; 8.9; 7.5 and 6.8 for $Pd(NH_3)_n(OH_2)_{4-n}^{+2}$ and $\log_{10}K_2$ for the second ethylenediamine complex $Pd(en)_2^{+2}$ is 18.4 but K_1 is too large to measure, because a strong perchloric acid sufficiently acidic to obtain partial dissociation of $Pd(en)(OH_2)_2^{+2}$ is too far removed from a genuine aqueous solution to allow the mass-action law to be used. The bidentate heterocyclic (rigid planar) 1,10-phenanthroline ligand[43] shows a much lower K_2 because of steric hindrance in $Pd(phen)_2^{+2}$ but $Pd(phen)(NH_3)_2^{+2}$ and the five-coordinate $Pd(phen)(OH)_3^-$ can be studied.

Bjerrum took up the concepts maximum N_{max} and characteristic N_{char} coordination number from Werner. In many instances, the formation curve shows a long plateau before increasing again, such as $N_{char} = 2$ and $N_{max} = 4$ for mercury(II). Thus, ammonia shows $\log_{10}K_n$ = 8.8; 8.7; 1.0 and 0.8 with a much greater discontinuity than Pd(II), and again, the two first constants are almost too large to be readily measured. In Section 6, we return to the more complicated problem of copper(II) having $N_{char} = 4$ and $N_{max} = 5$.

One of the motivations for studying ammonia complexes is that (before deuterium and oxygen isotopes were available) they offered one of the rare opportunities to get an idea about the constitution of aqua ions in solution. However, this does not always proceed as smoothly as in the case of nickel(II). Bjerrum found that $Zn(NH_3)_4^{+2}$ does not react further with aqueous ammonia, as confirmed by crystal structures containing this tetrahedral complex. However, the zinc(II) aqua ions in crystals are octahedral with $N = 6$ like magnesium(II) and nearly all 3d group aqua ions.

Such discrepancies may be quite frequent, especially among colorless closed-shell cations. Thus, ultra-violet spectroscopic evidence is obtained [44] for the silver(I) aqua ion being tetrahedral $Ag(OH_2)_4^+$ whereas concentrated ammonia [40] does not react with linear $Ag(NH_3)_2^+$. It may be added that Raman spectra [45] suggest the presence of $Ag(NH_3)_4^+$ in liquid ammonia, as well as $Ca(NH_3)_6^{+2}$ and $Zn(NH_3)_4^{+2}$.

It should not be neglected that studies of complex formation constants in *solution* provide information about another kind of *N* values than crystal structures. An interesting case is that Raman spectra of thallium(III) in strong hydrochloric acid indicate tetrahedral $TlCl_4^-$ but that $Co(NH_3)_6^{+3}$ and $Rh(NH_3)_6^{+3}$ immediately precipitate insoluble salts of $TlCl_6^{-3}$ like they do of $InCl_6^{-3}$ or of mixtures of lead(IV) and lead(II) forming [46] a purple compound $[Rh(NH_3)_6][PbCl_6]$ containing equal amounts of the well-known yellow $PbCl_6^{-2}$ and of $PbCl_6^{-4}$ of which the high *N* is surprising, in view of the amphoteric lead(II) species at high pH being $Pb(OH)_3^-$ to be compared with tetrahedral $Be(OH)_4^{-2}$.

4 Crystallographic Definition of Bonds

An ideal gas of geometrical points shows a probability density at a distance r from an arbitrary point proportional to $4\pi r^2\,dr$. It is convenient to divide an observed radial density P(r) of other nuclei (relative to a nucleus at origo) by this expression. In principle, such radial distributions can be obtained for liquids and for vitreous and amorphous solids by X-ray diffraction (where the strong maxima of electronic density exactly, or almost precisely, coincide with the nuclear positions) or by neutron diffraction. For gaseous molecules, such P(r) can be obtained by electron diffraction. It is important to realize [16,47] that crystal structures of solids constituted by translationally repeated unit cells involve two kinds of averaging: they can only provide the *time average* value of the *average content* of the unit cell. Other experimental techniques (such as optical and photo-electron spectra, nuclear magnetic resonance and Mössbauer effect) may legitimately suggest a lower local symmetry than obtained by crystallography, and they may disagree with each other according to their characteristic time-scale, which is of the order 10^{-13} s for visible absorption bands, and 10^{-17} s for photo-electron ionization.

Crystallographic studies provide two grades of internuclear distances R, between nuclei at special positions (frequently with a precision 0.005 to 0.002 Å) determined by geometric coefficients times the unit cell parameters, and between two nuclei which are both (or at least one) on general positions (typically 10 times less precise). It should be noted that R values with 4 or 5 decimals [48] usually are derived from micro-wave rotational spectra, especially of gaseous diatomic molecules, giving the time average of R^{-2}. The thermal vibrations at room temperature frequently have an amplitude of 0.05 to 0.1 Å which would only be decreased by a factor around 2 or 3 at the absolute zero.

If we plot the $P(r)/r^2$ for many instances of two nuclei Z_1 and Z_2 it is well-known that there are no *very* short internuclear distances observed. There is a sharp maximum at a characteristic value $r = R_{12}$. The peak is slightly asymmetric (it is much more frequent for an observed r to be 0.2 Å longer than R_{12} than to be 0.2 Å shorter) and its width is not much larger than the typical amplitude of thermal vibration at room temperature. This peak is what crystallographers call a chemical bond, and R_{12} is the typical bond-length.

For most choices of Z_1 and Z_2 there are hardly observed any r values up to a slope starting to grow up at $R_{12} + 1$ Å. The subsequent hilly landscape on the plot correspond to "Van der Waals contacts" and the oscillations attenuate for large r to a roughly constant line. Certain sulfur-sulfur and iodine-iodine weak interactions provide r values in the valley between the chemical bonds and the Van der Waals contacts, and Dunitz [49,50] discussed the relations between such phenomena and the transition path of a simple chemical reaction. Of course, there is not an absolute analogy between a hundred crystal structures available, involving a given triatomic group, and the amplitude of a single compound heated to an unrealistic high temperature, but one can still draw interesting conclusions. Thus, the many instances of linear, but asymmetric I_3^- anions with two differing R can be considered as steps in the reaction between I_2 and I^-, the symmetric anion (which seems to occur in nitromethane solution) representing the transition state of the exchange reaction. Traditionally, short hydrogen bonds between two oxygen atoms, or in FHF^-, are also considered to fall in the interval of R shorter than the Van der Waals contacts, because it used to be very difficult to determine the positions of the protons by X-ray diffraction.

Under equal circumstances, the X-ray diffraction pattern shows an intensity proportional to Z_1Z_2 and by far the larger part of the intensity is due to the electrons in inner shells. It is very difficult to obtain reliable values for the density of the valence electrons [51] and even in the very favorable case of diamond (where only a-third of the electrons are in inner shells) the electron density around each C^{+4} core is almost exactly spherical, and certainly not four sausages connecting the closest neighbor atoms. For group-theoretical reasons, the first deviation from spherical symmetry allowed [16] has octupole symmetry (proportional to xyz) but the strength of this deformation is only a few percent of the spherically symmetric density.

The question of valence-electron density has a specific significance [4] in the NaCl-type crystals NaF, KCl and RbBr and the CsCl-type CsI which are known, within experimental uncertainty, to show the X-ray diffraction pattern of the corresponding types with identical atoms. However, if it comes to look for differences of about one electron between the two kinds of atoms, present-day experimentation is rather far from the goal. [51]

If the N direct neighbor nuclei have the same Z and the same R, the crystallographic coordination number is N without discussion. If the Z values differ, even Werner did not argue that all six distances from chromium(III) in $Cr(NH_3)_4FBr^+$ are identical, only that the nuclei are situated on the Cartesian axes. If the R_{12} are similar to other instances, one still speaks of $N = 6$. However, this can become a matter of taste. The most recently refined crystal structure of LaF_3 shows 11 fluoride neighbors, but two of the distances are so much longer than the nine other (somewhat scattered) R values that there is no clear-cut answer to whether N is 9 or 11. Generally, R values at most 10% longer than the shortest value tend to be taken into account.

A special case of indeterminate N occurs [4] in the cubic type (representing the common modification of Fe, Cr, Mo, W, Nb, Ta and the alkaline metals) corresponding to CsCl with identical atoms. In the three cubic binary halide types CuCl ($N = 4$), NaCl ($N = 6$) and CsCl ($N = 8$) all atoms are on special positions, but the clear-cut distinction between cations and anions would permit no doubt about $N = 8$ in CsCl, whereas the six next-nearest neighbor nuclei in the metals mentioned have R values only 1.1547... times those to the eight nearest neighbors, and it may be argued that $N = 14 = 8 + 6$, slightly higher than $N = 12$ for the close-packed (cubic and hexagonal) types.

5 High *N* Values Without Available Electron-pairs

It seems [4] that the main (though not exclusive) parameter determining *N* in the case of identical neighbor nuclei is simply *relative atomic size*. When a central atom is very small (as in BF_3 and NO_3^-) equilateral $N = 3$ is obtained. P(V), S(VI), As(V) and Se(VI) in PO_4^{-3}, SO_4^{-2}, AsO_4^{-3}, SeO_4^{-2}, their protonated monomers, and oxygen-bridged pyrophosphate, metaphosphates, pyrosulfate, . . . all locally have tetrahedral $N = 4$ (which is only known from highly oxidizing nitrogen(V) compounds such as ONF_3 and NF_4^+). As predicted by Pauling [52] by extrapolation from $Sb(OH)_6^-$ and para-periodates with $N = 6$, the most stable xenon(VIII) complex is perxenate XeO_6^{-4}. Whereas CO_2 is linear with $N = 2$ and does not polymerize (like monomeric formaldehyde H_2CO does), the crystalline modifications quartz, cristobalite and tridymite of SiO_2 stable at normal pressure, as well as vitreous silica, all have locally tetrahedral $N = 4$, each oxygen atom bridging two silicon atoms. TiO_2 has approximately octahedral $N = 6$ in both rutile and anatase, whereas ZrO_2 and HfO_2 both crystallize in the low-symmetry type baddeleyite with $N = 7$. Both cassiterite SnO_2 and the plattnerite modification of PbO_2 crystallize in rutile type ($N = 6$). On the other hand, the cubic CaF_2 type ($N = 8$) is represented by CeO_2, PrO_2 and TbO_2 (readily loosing O_2 to form sub-stoichiometric PrO_{2-x} and TbO_{2-x}) and all nine dioxides from ThO_2 to CfO_2. As examples of the frequent gross non-stoichiometry of non-metallic inorganic compounds may be mentioned the fluorite-type of $Zr_{0.9}Y_{0.1}O_{1.95}$ used in Nernst lamps (because it conducts oxide ions between the vacant sites at red heat, and emits a brillant light when the electric current is increased) and the comparable $Zr_{0.9}Mg_{0.1}O_{1.9}$ and $Zr_{0.9}Ca_{0.1}O_{1.9}$ used as crucible materials. In analogy to the mineral thorianite, it is possible to prepare $Th_{1-x}La_xO_{2-0.5x}$ with x up to 0.54, remaining a statistically disordered fluorite lacking up to 13.5% of the oxygen nuclei, and still colorless. Such mixed oxides (with perceptibly broadened lines in the Debye powder diagram due to marginally different unit cells) can be prepared readily [53] by calcining the co-precipitated hydroxides obtained by adding ammonia to the mixed salt solutions.

Yttrium(III) and trivalent lanthanides (having the generic name "rare earths") have $N = 6$ for two sites (both differing strongly from octahedral symmetry) in the cubic C-type M_2O_3 and $N = 7$ in A-type La_2O_3 and La_2O_2S (the red cathodoluminescence in modern color television is emitted as spectral lines [27] of $4f^6$ europium(III) in $Y_{2-x}Eu_xO_2S$). It is remarkable to what extent [28,29,54] trivalent rare earths are indifferent to choice of $N = 7, 8, 9, 10, 11$ or 12 and toward selecting a given local symmetry. Unfortunately for the chemists, Euclid described polyhedra by their number of surfaces (of which a regular icosahedron have 20, but only 12 apices) which is irrelevant to chemists, and somewhat arbitrary when slightly differing R values provide a distorted polyhedron [54]. Werner believed that not only tetrahedral and octahedral coordination is preferred (as also seen [8]) from the Madelung energy) but also that the three other regular polyhedra are stable, and he assumed that the molybdenum(IV) complex $Mo(CN)_8^{-4}$ is cubal. Actually, cubal $N = 8$ is exceedingly rare in monomeric species (but stabilized by Madelung energy in CaF_2 type crystals) as can already be rationalized by the argument [55, 56]) that a tetragonal (Archimedean) anti-prism with the lower square of a cube turned 45° in its plane decreases the ligand-ligand repulsion in any model monotonically decreasing with X—X distance. This does not prevent that other symmetries of $N = 8$ may be even more stable.

It might be argued that the tolerance of rare earths to a variety of N between 6 and 12 is due to essentially electrovalent bonding. It seems rather to be a question of large cation radii. The same variability is observed in crystals containing thorium(IV), calcium(II), strontium(II) and barium(II), whereas the smaller magnesium(II) sticks to octahedral $N = 6$ in nearly all cases, including the aqua ion, rutile-type MgF_2 and NaCl-type MgO, and the even smaller beryllium(II) to tetrahedral $N = 4$. For comparison with the fluorite-type mixed oxides with anion deficit mentioned above, it may be mentioned that yttrofluorite $Ca_{1-x}Y_xF_{2+x}$ and UO_{2+x} show anion excess. This propensity toward gross non-stoichiometry in fluorites may be connected with the CaF_2 type systematically lacking half of the cations in the CsCl type.

Not all observed N values above 4 can be considered irrelevant electrovalent deviations from the Lewis paradigm; some correspond to pretty covalent bonding. By the way, it cannot be concluded that the gas SF_6 is much more covalent than the Si-F bonds in the isoelectronic SiF_6^{-2}. Magnus [57]) pointed out that the factual observation that anhydrous halides fall in two categories (one with rather low freezing- and boiling points, not conducting electricity in the liquid state, . . . and another with high melting points showing ionic conductivity in the molten state) is not so much a black-and-white separation between covalent and electrovalent compounds, as a question of mobile molecules having $N = z$ and the high-melting salts N higher than the oxidation state z, as when comparing $N = 6$ for NaF, MgI_2 and AlF_3 with the gaseous molecules SiF_4, PF_5 and SF_6. Sometimes, N may be decreased by evaporation, as when crystalline $AlCl_3$ and $FeCl_3$ having $N = 6$ forms the gaseous dimer $Cl_2MCl_2MCl_2$ with tetrahedral $N = 4$. The corresponding iodine(III) and gold(III) chlorides have all 8 nuclei in the same plane, and already the solid consists of molecules. It is a general trend in series of homologous organic compounds that the boiling-point increases with the molecular weight (except the decrease, when fluorine substitutes for hydrogen). This tendency can be quite extreme in analogous inorganic molecular solids, WCl_6 boils 330° higher than WF_6. Since the boiling-point is essentially given as the ratio between the heat of evaporation and the entropy invrease in the gaseous state, it is generally argued that such strong changes are due to Van der Waals attractions between induced multipole moments on adjacent molecules, as seen in the strongly increasing boiling-point of the elemental halogens from F_2 to I_2. Sometimes, the effect may have the opposite sign of expected; with exception of LiI, all alkali-metal halides have lower vapor pressure between 10^{-3} and 1 atm than cesium iodide [27]). This might be related to the shorter R = 3.315 Å in the diatomic molecule [48]) than 3.95 Å in the crystal. By the way, such crystals may be dispersed in mass spectrometers [58]) showing a long series of oligomers $Cs_{n+1}I_n^+$ with n from 1 to above 60 leaving the ion-source.

Several cases of carbon are known with N higher than 4. When alkyl groups bridge beryllium or aluminium (and it is symptomatic that these atoms are rather small) $N = 5$. The molecules $Li_4(CH_3)_4$ shows $N = 6$, each carbon atom being bound to three atoms of the surface of the tetrahedral group Li_4 (like P_4) and to three terminal hydrogen atoms. The enigmatic CH_5^+ is important for mass-spectra and quantum chemists (much like H_3^+). The molecule $CRu_6(CO)_{17}$ shows $N = 6$ for the carbon atom in the center of the octahedral Ru_6 group, much like the NaCl-type carbides MC known both for closed-shell K values such as M = Ti, Zr, Hf and Th and other systems with d-like or f-like electrons or with delocalized conduction electrons such as M = V, Ta, U, Np and Pu. The amber-yellow non-metallic beryllium carbide Be_2C crystallizes in CaF_2 type, and hence conventional tetrahedral $N = 4$ for Be(II) but cubal $N = 8$ for C(—IV) having only half the number of valence electrons demanded by the Lewis paradigm (whereas SiC is well-behaved).

It has become customary to speak of these high N values for carbon as electron-deficient, a characteristic found in most boranes and carboranes [59] whereas other N values (as found in PF_5 and SF_6) are called hyper-valent. This obedience to the Lewis paradigm reminds one about meteorological weather predictions, to which the speaker adds that a considerable amplitude of hyperthermic and hypothermic deviations are expected, as well as xeric and hydrophilic anomalies.

The inorganic chemist tends to consider organo-metallic compounds (having carbon directly bound to atoms of metallic elements) differently from the recent fashionable text-books. The unidentate isoelectronic ligands CO, CN^- and the related isonitriles [60] tend to have moderate N such as 4, 5 and 6 (providing an enormous number of diamagnetic d^6 complexes) though CO and PF_3 enhance reduction to negative or zero oxidation states having no similarity to the corresponding gaseous atom [8]. Within the specific domain of organo-metallic compounds of the d-groups, the Lewis paradigm is replaced by the 18-electron rule that systems having an even number q of d-like electrons are expected to show $N = 9 - 1/2q$. There is no doubt that this rule is not reliable in general for d-group chemistry. Grim exceptions to the expectation of even q are Cr(III), Mn(II) and Fe(III) representing the most frequent oxidation states of their elements, and it is distinctly not true that octahedral complexes of Cr(III) and Mn(IV) are readily reduced to Cr(0) and Mn(I); nor of Ni(II) readily oxidized to Ni(IV) or $Cu(NH_3)_5^{+2}$ readily oxidized to Cu(III). It is not even true for most octahedral iron(III) and cobalt(II) complexes that they are readily reduced or oxidized, respectively.

However, as far goes d-group organo-metallic compounds, the 18-electron rule works better than the Lewis paradigm does in general inorganic chemistry. It is not easy to find representatives of all N values predicted. The best cases $N = 9$ without d-like electrons ReH_9^{-2} and WL_3H_6 (in the following, we write L for triphenyl-phosphine $P(C_6H_5)_3$) are not strictly organo-metallic, but the phenomenon [61] of "inorganic symbiosis" makes hydride and phosphines readily co-existing with carbon-bound ligands. $N = 8$ and q = 2 are known from molybdenum(IV) and tungsten(IV) in $M(CN)_8^{-4}$ though these species can be oxidized to M(V) in $M(CN)_8^{-3}$. $N = 7$ and q = 4 is represented by quite a lot of Mo(II) and W(II) organo-metallic complexes, and by IrL_2H_5 which has the rather unexpected feature of being one of the two best characterized iridium(V) complexes, the other being IrF_6^-. There is a plethora of $N = 6$ and q = 6 including IrL_3H_3 and the nickel(IV) dithiocarbamate [62]

$Ni(S_2CN(C_4H_9)_2)_3^+$. The first large-scale series of exceptions occur for q = 8 very frequently having quadratic $N = 4$ such as $M(CN)_4^{-2}$ for M = Ni, Pd and Pt. Even when $N = 5$, it is sometimes tetragonal-pyramidal $Ni(CN)_5^{-3}$ (like green d^7 $Co(CN)_5^{-3}$ and d^9 $Cu(NH_3)_5^{+2}$) rather than the more conventional trigonal-bipyramidal species isosteric and isoelectronic with $Cr(CO)_5^{-2}$, $Mn(CO)_5^-$ and $Fe(CO)_5$. The majority of platinum(II) complexes are quadratic, including [8] yellow $Cl_2Pt(SnCl_3)_2^{-2}$ but dark-red $Pt(SnCl_3)_5^{-3}$ is trigonal-bipyramidal. If the d-shell is completed, the 18-electron rule suggests tetrahedral $N = 4$ as well-known from $Mn(CO)_4^{-3}$, $Fe(CO)_4^{-2}$, $Co(CO)_4^-$, $Ni(CO)_4$, the palladium(0) complex $Pd(CN)_4^{-4}$ and from $Os(PF_3)_4^{-2}$, $Ir(PF_3)_4^-$ and $Pt(PF_3)_4$. However, PtL_3 is known to be monomeric with $N = 3$, and linear $N = 2$ is frequent for Cu(I), Ag(I), Hg(II) and even $Tl(CH_3)_2^+$ isoelectronic with $Hg(NH_3)_2^{+2}$ and is almost universal for Au(I). Excepting heavy steric constraint (as in $(H_3C)_3CBeC(CH_3)_3$) cases of $N = 2$ are clear-cut evidence for covalent effects limiting N to a lower value than determined by relative atomic size [4] alone. Pauling suggested that CoF_2 and $KCoF_3$ have $N = 6$ but CoX_4^{-2} with the three heavier halides $N = 4$ because each fluoride ligand donates much less electronic density to Co(II) than X^- does. There is much to be said for this qualitative approach [8] though much spectroscopic evidence from the transitions in the partly filled d-shell described by "ligand field" theory indicates fractional charges of the central atom typically between +1 and +2 (and not necessarily increasing with z) whereas Pauling expressed the feeling that they occur in the interval between zero and +1. An extreme case of the high N for fluorides can be seen in HgF_2 with $N = 8$ like CdF_2 and BaF_2 whereas mercury(II) with most other ligands [40,41] reluctantly increases N from 2 to 4. The fiasco for the 18-electron rule for q = 10 is accompanied by an unexpected oscillation of N as function of z in the isoelectronic series $K = 28$, 46 and 78. Thus, $N = 4$ is nearly universal in Mn(—III), Fe(—II), Co(—I) and Ni(0) and shows a dip toward $N = 2$ in the single case Cu(I). The two next, Zn(II) and Ga(III), have more frequently $N = 6$ than 4. The preference shifts slowly back from 6 to 4 in Ge(IV), As(V), Se(VI) and Br(VII).

After this survey of the 18-electron rule applied to complexes of unidentate ligands, we may discuss an almost metaphysical denticity introduced in sandwich complexes, where a certain number of carbon atoms (or a mixture of boron and carbon atoms [59] in carborane ligands) situated in a regular polygon are bound at equal R to the d-group atom. Thus, an effective $N^* = 3$ is ascribed to cyclopentadienide $C_5H_5^-$, benzene C_6H_6 and the tropylium cation $C_7H_7^+$. One origin is mixed complexes such as $(C_6H_6)Cr(CO)_3$ clearly having the crystallographic $N = 9$ though the six Cr–C distances to the benzene ligand are considerably longer (2.22 Å) than the three R = 1.84 Å to the unidentate ligands [63]. These distances differ perceptibly from 2.14 Å in $Cr(C_6H_6)_2$ and 1.92 Å in $Cr(CO)_6$. The crystal structure of $Cr(C_6H_6)_2$ has now been refined to a point where it is certain that $N = 12$ shows no traces of Kekulé alternation, which might have provided an argument for $N^* = 6$. By the way, even in the latter description, the 18-electron rule is not obeyed by the chromium(I) cations readily obtained with mild oxidants. Actually, $Cr(C_6H_6)_2^+$ and its substituted analogs [63] form the very large majority of all well-characterized Cr(I) compounds.

The situation surrounding $N^* = 6$ in $M(C_5H_5)_2$ and $M(C_5H_5)_2^+$ is far more confusing from a group-theoretical point of view because 3 and 5 have no common prime-factor (and actually both are prime-numbers). A major reason for this ackward classification is that the S values observed (1/2 for d^5 and d^7; 0 for d^6 and 1 for d^8) are the same as obtained in an octahedral chromophore with a large energy difference between the two sub-shells, the lower sub-shell having angular functions proportional to (xy), (xz) and (yz). It has been verified by photo-electron spectra [64,65] of the gaseous ferrocene molecule that the two orbitals ($x^2 - y^2$) and (xy) (having the same energy by group-theoretical necessity) turn out to have almost the same energy as ($3z^2 - r^2$) (rotationally symmetric around the 5-fold axis) whereas the two strongly anti-bonding orbitals (xz) and (yz) contain one electron in $Co(C_5H_5)_2$ and $Ni(C_5H_5)_2^+$ and two electrons in $Ni(C_5H_5)_2$ corresponding [66] to strongly reducing character of cobaltocene and nickelocene. This state of affairs suffices to produce the same S values as in low-spin octahedral complexes, though for an entirely different reason. It might be argued that the three ligands mentioned have six bonding π electrons in the Hückel model of planar aromatic systems, but such a justification for $N^* = 3$ of each ligand is readily extended to oxide ligands in $MoOCl_5^{-2}$ and nitride ligands in NSF_3 (which has been said to have $N^* = 6$ rather than $N = 4$) and the advantage of $N^* = 6$ (and not the crystallographic $N = 10$) in cyclopentadienide sandwiches does not seem convincing.

Uranium(IV) forms a large number of $(C_5H_5)_3UX$ with $N = 16$. The same N is found in uranocene $U(C_8H_8)_2$. When comparing with other cyclo-octatetraenide sandwich complexes [67] such as $Ce(C_8H_8)_2^-$, $Th(C_8H_8)_2$ and $Am(C_8H_8)^-$ it is not evident that the 4f or 5f orbitals play a decisive rôle [65]. This conclusion is corroborated by recent evidence [68] for other U(IV) compounds. The covalent bonding in such cases is mainly taken care of by empty orbitals in the L.C.A.O. model, 5d and 6s in the lanthanides [27,28] and 6d and 7s in the 5f group [18,30,31].

We do not have the space here to discuss the complicated question [65,69] of bonding between d-group atoms in organo-metallic compounds. Two simple extremes [8] are $V(CO)_6$ isoelectronic with $Fe(CN)_6^{-3}$ (both having $S = 1/2$ and being exceptions to the 18-electron rule) not known to dimerize, whereas the $Mn(CO)_5$ group has many of the characteristics of a pseudohalogen (a polyatomic group G^- able to dimerize to GG with loss of two electrons, like a large number of RS^- oxidized to RSSR). The yellow, diamagnetic (or anti-ferromagnetic?) $(OC)_5MnMn(CO)_5$ is strictly isoelectronic with Adamson's purple dimer $(NC)_5CoCo(CN)_5^{-6}$. In this sense, the green tetragonal-pyramidal $Co(CN)_5^{-3}$ with $S = 1/2$ is isoelectronic with the unknown $Mn(CO)_5$ and it reacts with benzyl iodide $C_6H_5CH_2I$ to form equal amounts of the two cobalt(III) complexes $Co(CN)_5I^{-3}$ and $C_6H_5CH_2Co(CN)_5^{-3}$ (and it reduces H_2 to the Co(III) hydride complex $HCo(CN)_5^{-3}$ isoelectronic with $HMn(CO)_5$) in a way reminiscent of Grignard reactions.

As recently discussed in more detail [4] the regular octahedral $N = 6$ in NaCl-type crystals is compatible with many kinds of chemical bonding. With a minor reservation [8,16] in the cases of AgCl and AgBr, it may very well be that the binary halides are almost exclusively electrovalent. Already the oxides MgO, CaO, MnO, CoO, NiO, SrO, CdO, BaO and EuO show some sign of beginning covalent contributions. This becomes much more pronounced in the MN, MP, MAs, MSb and MBi formed by yttrium and most of the lanthanides, and reaches a climax in the MC discussed above. Seen from the point of view of the Lewis paradigm, the oxides, nitrides and carbides present the most serious problem [2]. There are at most 8 outer electrons available in each of these anions, but 12 would be needed for six two-electron bonds.

It is also argued in the L.C.A.O. model that only the 1s orbital is available for bonding in hydrogen compounds, because the 2s and 2p orbitals have a 4 times smaller ionization energy than 1s in the neutral atom. Hence, it was early felt to be a deviation from Lewis' paradigm that FHF^- is a stable species with $N = 2$ for hydrogen(I). Similar short, symmetric hydrogen bonds are found in some crystal structures containing $H_2OHOH_2^+$ but Giguère [70,71] has shown that H_3O^+ and OH^- do not form symmetric hydrogen bonds in aqueous solutions. However, their great mobility is related to their exceedingly short life-time of a few picoseconds, and $H_9O_4^+$ is not a collective species but is produced by comparatively weak hydrogen bonds between OH_3^+ and three OH_2 molecules. By the way, vibrational spectra [72] of aqueous hydrofluoric acid shows that the species having pK close to 3 is an intimate ion-pair $OH_3^+F^-$ and does not involve diatomic HF. This result makes a lot of text-book explanations of the lower acidity of HF than of HCl rather superfluous. Actually, the pK close to −9 for HCl derived from Hammett functions, corresponding to almost complete reaction with water (forming the much weaker acid H_3O^+) is confirmed [73] by the very moderate vapor pressure of HCl in aqueous solution below 2 molar concentration. Evidence has been reviewed [73] for a certain kind of protonation of anhydrous liquid hydrogen fluoride under exceptional circumstances, but the detailed constitution of $H_{n+1}F_n^+$ is not known.

It has attracted less attention [3,4] that both linear and bent hydride bridges occur with $N = 2$ for H(—I). The classical case of an "electron-deficient" compound is diborane $H_2BH_2BH_2$ isosteric with Al_2Cl_6. Another colorless molecule with bent hydride bridges is $H_2BH_2BeH_2BH_2$. Volatile $Zr(H_3BH)_4$ with $N = 12$ for Zr(IV) and related cases for Hf(IV) and U(IV) were reviewed [4]. All five alkali-metals form NaCl-type MH with $N = 6$ for H(—I). Mixed oxides of perovskite type are numerous [4,28] but this type is also represented by $BaLiH_3$ and $EuLiH_3$ having the common $N = 6$ for lithium(I), and each hydride bound to two Li(I) on one Cartesian axis and to four M(II) situated on the two other axes.

6 Aqua Ions and Related Problems

Water is the most important solvent for inorganic chemistry, and since the proposal by Arrhenius of extensive ionic dissociation, there has been a continuous struggle to characterize the aqua ions. As far goes cations, the hydration energy [8,28,33,74] from gaseous ions M^{+z} is usually proportional to z^2 but it can only be described by the stabilization $(z^2/2r) \cdot 14.4$ eV/Å by a perfect dielectric (the reciprocal dielectric constant 0.013 of water is so small that it may equally well be neglected at this level of precision) if r is about 0.83 Å larger than the Goldschmidt ionic radius, as first pointed by Latimer [75]. In the other hand, the four halide X^- show quite closely the hydration energy expected for a perfect dielectric. This discrepancy cannot be remedied by choosing another basis for ionic radii than Goldschmidt. As first pointed out by Fajans in 1919, the Madelung constant is close to 1.75 for binary MX, and the hydration should bring a minimum stabilization corresponding to a constant 2 (obtained when M^+ and X^- have the same radius). This difference would provide a highly exothermic dissolution [76]. As a matter of fact, the lithium salts and cesium fluoride dissolve with a much smaller evolution of heat, and the other halides hardly any temperature change at all, with CsI dissolving endothermally. This spherically symmetric model of hydration energy hardly leaves any possibility of evaluating a definite coordination number *N*.

The characterization of aqua ions as Werner complexes starting with the instantaneous equilibrium (with pK close to 5.7, one unit higher than of acetic acid) between brick-red $Co(NH_3)_5OH_2^{+3}$ and bluish red $Co(NH_3)_5OH^{+2}$ was extended by Niels Bjerrum 1909 to $Cr(OH_2)_6^{+3}$ using visible absorption spectra. The half-life of the Cr–O bonds is known to be about a day, and much longer for $Rh(OH_2)_6^{+3}$ and the more recently [77] isolated $Ir(OH_2)_6^{+3}$ showing reaction rates comparable to many organic compounds. Like alums serve to indicate the presence of $M(OH_2)_6^{+3}$, schönites (Tutton salts) such as $K_2[M(OH_2)_6](SO_4)_2$ serve to recognize M(II) hexa-aqua ions.

The status of well-defined *N*, kinetic life-time and stereochemical rigidity differ tremendously among various aqua ions [38,78,79] and in particular, many questions remain unresolved among the colorless closed-shell ions. Tetrahedral $Be(OH_2)_4^{+2}$ and octahedral Mg(II), Al(III), Zn(II), Ga(III) and probably Cd(II) are well-established. Comparison of absorption spectra of solutions with reflection or transmission spectra of crystals with known structure has allowed identification of octahedral Ti(III), V(II), V(III), Cr(III), Mn(II), Fe(II), Fe(III), Co(II), Co(III), Ni(II), as well as comparative "ligand field" arguments for Ru(II), Ru(III), Rh(III) and Ir(III) aqua ions. The situation is more complicated for Cr(II), Mn(III) and Cu(II) aqua ions with a problem in common. There is no doubt that Tutton salts are known, containing $Cr(OH_2)_6^{+2}$ and $Cu(OH_2)_6^{+2}$. However, their spectra are slightly differing from the aqueous solution, and both series of spectra do not suggest cubic symmetry with six identical distances R but rather four short R in a plane. A further problem [74,80] is that blue $Cu(NH_3)_4(OH_2)^{+2}$ seems to be tetragonal-pyramidal like $Cu(NH_3)_5^{+2}$ whereas anhydrous $Cu(NH_3)_4^{+2}$ syncrystallized as traces in $[Pt(NH_3)_4](CH_3C_6H_4SO_3)_2$ is pink. Though $[Ni(NH_3)_6]I_2$ is a well-behaved cubic crystal, the corresponding copper(II) compound contains statistically disordered $Cu(NH_3)_5^{+2}$. The conclusions for the aqua ion are two-fold: the equatorial rôle of short R is very rapidly taken over by the fifth (and sixth?) water molecule, also providing a water exchange in less than 10^{-8} s to be compared [79] with 10^{-4} s for $Ni(OH_2)_6^{+2}$, and even the ligands (if there are two) on the axis perpendicular on the instantaneous equatorial plane (with 4 short R) may not be equivalent. The high band intensities [74] of Cu(II) compared with Ni(II) having similar sets of ligands (and $S = 1$) suggest a systematically distorted, instantaneous picture [16]. This situation is even more pronounced in palladium(II) complexes [42,74] curiously enough having almost exactly twice the transition energies of the corresponding copper(II) complex. It is likely that $Pd(OH_2)_4^{+2}$ is straightforward quadratic [81] but not absolutely excluded that it has a weakly bound, fifth water molecule perpendicular on the $Pd(II)O_4$ plane.

The neutron diffraction of aqueous solutions containing different stable isotopes of a given element, and allowing comparison of deuterium and normal hydrogen, recently have allowed new determinations of average values of *N* and R in aqua ions. Thus, calcium chloride solutions [82] going from 1 to 3.9 molar decrease *N* from 10 to 6. This result does not imply that all five integers from 10 to 6 are strongly represented, nor does it give any significant information about the angular distribution and its symmetry. The result is rather similar to the strong variation of *N* for calcium(II) in crystals. There is a very longstanding controversy as to whether $N = 9$ or 8 for aqua ions of yttrium(III) and the trivalent lanthanides. It seems [29] as if the Z values above 63 (as well as yttrium having an ionic radius corresponding to 66.5) tend toward 8 in stronger solutions, in spite of $[M(OH_2)_9](BrO_3)_3$ and $[M(OH_2)_9](C_2H_5OSO_3)_3$ being known for all M in two series of isotypic salts. Recent sophisticated neutron diffraction experiments [83] on 2.85 molar neodymium(III) chloride gave evidence for comparable amounts of $N = 9$ and 8 being present. One of the interesting features is the detection of the 17 protons at closest distance from Nd(III).

It seems likely at present that aqua ions fall roughly in three categories. When N is 6 or lower values, they are chemical species in the same sense as the ammonia complexes studied by Werner and Jannik Bjerrum, though we saw in Section 3 that tetrahedral $Ag(OH_2)_4^+$ is not analogous to linear $Ag(NH_3)_2^+$ in aqueous ammonia, nor octahedral $Zn(OH_2)_6^{+2}$ to tetrahedral $Zn(NH_3)_4^{+2}$. There may be hanging a few more surprises around. In view of $N = 8$ in crystalline HgF_2 it would not be too surprising if mercury(II) aqua ions have $N = 6$ but show strong distortions on an instantaneous picture in the opposite direction of copper(II) aqua ions, i.e. two short R on one Cartesian axis, and four long R on the two other axes.

The second category of aqua ions corresponds to a mixture of differing N values (and symmetries) with almost the same free energy, readily modified by changing salt concentration, temperature (and perhaps even hydrostatic pressure). The rather indifferent choice between $N = 7$, 8, 9 and 10 may be characteristic for calcium(II), the rare earths and probably also for thorium(IV). The third category are cations with so large radii and so low positive charge (probably K^+, Rb^+, Cs^+ and Ba^{+2}, perhaps even Tl^+) that the nearest water molecules are so relatively weakly perturbed and oriented by the cation that the situation is similar to the second layer of water molecules around $Co(NH_3)_6^{+3}$ or $Cr(OH_2)_6^{+3}$, or even $N(CH_3)_4^+$.

One of the intrinsic properties of aqua ions is that they may deprotonate to hydroxo complexes in solution (or precipitate hydroxides) and eventually deprotonate further to oxo complexes. To the first approximation [4,38] this deprotonation proceeds according to three parameters: pH increasing from -1 to 15 (roughly the range in which the solution is still genuinely aqueous), the oxidation state z increasing from $+1$ to $+8$, and the ionic radius r_{ion} decreasing. A nice "physical" theory would be that the important parameter is the ratio (z/r_{ion}). This is approximately true for electrovalent series such as the alkaline-earth M(II) and the rare-earth M(III). When this ratio is sufficiently high, OsO_4 and ClO_4^- do not protonate perceptibly, $HOMnO_3$ and $HOReO_3$ have slightly negative pK, etc.

However, it is a quite striking fourth parameter that oxidizing aqua ions under equal circumstances are more acidic (have lower pK). Thus, iron(III) is distinctly more acidic than aluminium(III) and chromium(III), copper(II) more acidic than nickel(II) and zinc(II), and quite excessive acidity is observed for mercury(II) [41], palladium(II) [42,81] and thallium(III) aqua ions, compared with the much smaller beryllium(II) and aluminium(III). This tendency takes extreme proportions [84] in gold(III) complexes.

The deprotonation to hydroxo and oxo complexes modifies the more general theory of hydration energy [8,10,28,33] which almost allows the prediction of what z values are stable (e.g. toward evolution of O_2 by not being too oxidizing, and toward H_2 by not being too reducing) among the aqua ions, connecting the question with the variation of ionization energies of monatomic entities [8,24]. In this way, the chemical behavior of the 5g-group elements [32] between $Z = 121$ and 138, dwi-lead $Z = 164$ with palladium-like chemistry [85] and the Z values between 104 and 120, can be discussed where it seems [86] that eka-francium is readily oxidized from 119(I) to 119(III) or 119(V), which might already be true for francium too [10]. The concept of electronegativity has ramified to a large extent [8,87] and recently, Lackner and Zweig [88] suggested that the expected chemical properties of systems containing unsaturated quarks [12,89] are determined essentially by the Mulliken electronegativity of the corresponding monatomic entity. A closer analysis [90] shows that this is not a sufficient parameter, that well-defined counter-examples are available [24,28] in the 4f group, and that it is more appropriate to extrapolated chemical behavior of electronic densities surrounding tiny nuclei or quarklei [91] with charges $(Z + 1/3)$ and $(Z + 2/3)$ intercalating "new elements" in the Periodic Table. Fairbank and his group at Stanford University maintain that metallic niobium contains one unsaturated quark per 10^{20} units of atomic weight (that is 6000 per g).

Perhaps very fortunately, inorganic chemistry does not have one universal well-ordered deductive theory. An interesting case of the several complementary net-works of concepts used when rationalizing chemical facts can be found in the book [92] by Williams. A mathematician may think about such a book as written in the language "doublespeak" of Orwell's novel "1984". However, chemistry is much more inductive [10] than deductive, the facts are stubborn, and we do not yet know the surprises of tomorrow.

7 Special Extensions of the Lewis Paradigm: the Hybridization and the Kimball Models

The d-group chemist shows an intrinsic preference [93] for molecular orbital (M.O.) rather than valence-bond (V.B.) treatment. One root of this choice is that the discrete energy levels of monatomic entities [22,23,87] are correctly classified by the preponderant electron configuration [8] in spite of quite strong correlation effects modifying the total wave-function (when 4 or more electrons are present) and also explain optical transitions, both of the kind described by "ligand field" theory [8,16,94] and electron-transfer spectra [94,95]. This description is closely related to the principle of Franck and Condon [96] that the nuclei do not have the time to move during the quantum jump of the many-electron system. This principle turns out to be equally important for photo-electron spectra of gaseous molecules [26,64,65,97–99] (usually induced by 21.2 or 40.8 eV ultra-violet photons) and of solids [21,100–104] (usually induced by 1253.6 or 1486.6 eV mono-energetic X-rays).

The V.B. treatment may be useful for atoms separated by a large distance, e.g. for anti-ferromagnetic coupling [16]. The connection with the Lewis paradigm stems from the description of H_2 by Heitler and London. The general disadvantage [93] for V.B. treatment of the groundstate of oligo-atomic species is the enormous number of "resonance structures" most of which redistribute electric charges between the individual atoms involved. The beginning of this problem can be seen in the M.O. configuration $(\sigma_g)^2$ of two hydrogen atoms at large R of which the squared amplitude corresponds to 50% (H^0H^0) and 25% of each of the V.B. structures (H^-H^+) and (H^+H^-). This statement is, at the same time, also a criticism of M.O. treatment of systems having large R (as in anti-ferromagnetic situation) since the groundstate of two highly separated hydrogen atoms corresponds to 50 to 51% squared amplitude of $(\sigma_g)^2$ accompanied by 50 to 49% $(\sigma_u)^2$. The low-lying dissociate triplet ($S = 1$) state corresponds to the well-defined M.O. configuration $\sigma_g\sigma_u$. The V.B. treatment is rarely applied to excited states.

Anyhow, it is possible to concentrate attention on a predominant V.B. structure such as H^0H^0 (the two S values 0 and 1 obtained by Hund vector-coupling highly differing in energy) and to ameliorate the bonding orbital on each atom. One purpose is to rationalize the angular selective bonding. Though vibrational spectra show that angular bending usually needs much less energy than radial stretching of a polyatomic species, and though low bond angles such as 60° occur in cyclopropane and P_4 it is also striking how similar the bond angles are in a penta-atomic molecule consisting of a carbon atom bound to various hydrogen and halogen atoms to those of regular tetrahedral CX_4. The ameliorated orbital to be used in the predominant V.B. is obtained in the hybridization model of Pauling as a linear combination of orbitals with differing l values centered on a given atom. A simple case is the general expression for the mixture of s and p angular functions

$$(1 - \alpha^2 - \beta^2 - \gamma^2)^{1/2} + [(\alpha x/r) + (\beta y/r) + (\gamma z/r)]\sqrt{3} \qquad (2)$$

which are all four mutually orthogonal. It is well-known that the three p functions have many properties similar to vectors, whereas the first part of Eq. (2) is a scalar (lacking angular dependence). Not only the three p angular functions have the same geometrical shape, but the sum of the three last parts of Eq. (2) has also the same shape, turned in space according to the three coefficients α, β and γ (which can be chosen arbitrarily on the condition that the scalar square-root does not become imaginary).

A closer analysis [2)] shows that the hybridization model is a fair approximation for 2s and 2p orbitals of the elements from beryllium to fluorine. Their atomic core has only two electrons located quite close to the nucleus, and the radial functions 2s and 2p are sufficiently similar to be put outside a parenthesis to be multiplied by Eq. (2). Of course, it is not a mathematical requirement that the hybridized orbital (to be used in a V.B. treatment) can be approximately factorized in a radial and an angular part, but all the appealing characteristics of Eq. (2) disappear, if such a factorization is not a reasonable approximation. This is not the only striking difference between one-digit and two-digit Z values. Starting with sodium ($Z = 11$) all neutral atoms [21, 87)] have higher correlation energy (expressing the imperfection of the optimized Hartree-Fock function of a single configuration) than their first ionization energy. The "valence-state" energy of a carbon atom in an aliphatic compound is connected with the excited configuration $1s^2 2s 2p^3$ (several states, not only the lowest with $S = 2$) and increases with the scalar coefficient in Eq. (2). However, since the pioneer work of Turner [97)] and Price [99)] the ionization energies I of penultimate M.O. have become experimentally available, and not only does methane show $I = 14$ eV of three degenerate M.O. and $I = 23$ eV of one M.O. in close analogy with the neon atom having $I(2p) = 21.6$ eV and $I(2s) = 48.5$ eV, but all gaseous molecules containing carbon, nitrogen, oxygen and fluorine show deep-lying 2s-like M.O. What is perhaps more surprising [103] is that hydrocarbons containing n carbon atoms show n such penultimate M.O. (two in C_2H_6, six in C_6H_6 with four I values corresponding a fully occupied Hückel hexagon, and five in $C(CH_3)_4$ with three I = 18, 22 and 25 eV among which the middle correspond to three M.O. as expected from group-theory). In a sense, these results insinuate anti-bonding effects between 2s-like orbitals, as if the carbon atom remained in its lowest configuration $1s^22s^22p^2$.

The hybridization model is subordinated to the Lewis paradigm in the sense that N bonds have to be provided with N orthogonal orbitals suitable for the predominant V.B. with adjacent atoms. This requirement is entirely alien to the L.C.A.O. model of M.O. allowing the delocalization coefficients to be highly different for the two d-like, three p-like and one s-like M.O. of octahedral MX_6 having two, one and no angular nodes [16]. The quest for d-orbitals to include in the hybridization model applied to $N = 5$ and 6 is not successful [2]. Either the 3d radial function is far more extended than the 3s and 3p functions in PF_5 and SF_6, or the 4s and 4p orbitals have much greater average radii than the 3d orbitals in $Co(NH_3)_6^{+3}$ and NiF_6^{-2}. The last attempt to rescue the hybridization model applied to cobalt(III) complexes, which is known to the writer, also had serious problems [105].

The hybridization model was a refreshing innovation, when it was made by Pauling 1931, but the commentators in text-books have gone very far along a sterile scholastic desert trail. The main field of use today is describing bond angles, but the conclusions are a posteriori. It would be easy to construct a programme for a pocket computer to tell what you are supposed to say when seeing a given chromophore. You have the choice between "sp" and "sd" for linear XMX, your electronic parrot says "sp^3d^3f" for cubal MX_8, and though the conditioned reflex is "sp^3" when seeing a regular tetrahedron, a more sophisticated "sd^3" may be legitimate for permanganate. The trigonal MX_9 known from ReH_9^{-2} and $Nd(OH_2)_9^{+3}$ as well as anhydrous $LaCl_3$ inspires the comment "sp^3d^5". Actually, the stereochemistry predicted by the 18-electron rule for 2q d-like electrons can be labelled "sp^3d^{5-q}".

Gillespie [106] chose a closer link to the Lewis paradigm in describing bond angles as resulting from the mutual repulsion of electron-pairs of three kinds. Under normal circumstances, lone-pairs have the greatest demand for spatial angle on a spherical surface with the nucleus at center. The next-largest demands are for electron-pairs involved in bonding to elements with comparable electronegativity, and the smallest demand for electron-pairs connecting elements such as fluorine. Thus, the bond angle 103° in NF_3 is smaller than in NH_3 but larger than 96° in AsF_3. This does not prevent that the bond-angles in PH_3 and AsH_3 are much closer to 90°. The idea of lone-pairs originated with the resolution in 1900 of sulfonium cations with three substituents $RR'R''S^+$ by Pope into optically active enantiomers, as if they were tetrahedral with an invisible substituent. The Gillespie lone-pairs can hardly be detected in d-group compounds. The situation is very mixed in the post-transition-group central atoms such as As(III), Se(IV) and Br(V) with $K = 30$; In(I), Sn(II), Sb(III), Te(IV), I(V) and Xe(VI) with $K = 48$; and Au(—I), Tl(I), Pb(II), Bi(III) and Po(IV) with $K = 80$. At one side, they may occur in cubic crystals such as the K_2PtCl_6-type Cs_2SeCl_6 and Cs_2TeCl_6, the $SrTiO_3$ (perovskite)-type $CsPbCl_3$, the NaCl-type PbS, PbSe and PbTe, and the CsCl-type (non-metallic) CsAu. On the other side, they provide some of the most striking examples of lone-pairs replacing a ligand, such as pyramidal SeO_3^{-2}, BrO_3^-, $SnCl_3^-$, $SbCl_3$, $TeCl_3^+$, IO_3^-, XeO_3 and presumably also $Pb(OH)_3^-$; and tetragonal-pyramidal TeF_5^-, IF_5 and XeF_5^+ looking exactly as an octahedral complex lacking one ligand, what is also true for $OXeF_4$. This discrepancy may be related to the crystallographic restriction to time-average densities (Section 4). Arguments in this direction [16] would be low-symmetry TlI becoming CsCl-type (what TlBr always is) by heating, and the exceptionally early optical absorption of PbS may be connected with Pb(II) on an instantaneous picture tending to be situated at the eight equivalent places on the ternary axes, at some distance from the center of the unit cell. It may also be noted that $TeCl_6^{-2}$ provides a considerably larger cubic unit cell parameter than $SnCl_6^{-2}$. In this connection, it must be remembered that the inversion frequency of gaseous ammonia is very high; if pyramidal NH_3 is considered on a time-scale longer than 10^{-9} s, the average picture has a plane of symmetry, and the position of the nitrogen nucleus on an axis perpendicular on this plane of symmetry containing the three protons shows a camel-back-shaped double maximum with an intervening low value in the plane. This makes a difference from BF_3 but in both cases, the time-average exemplifies the point-group D_{3h}.

Like implicitly done by Lewis, Gillespie counts systematically the *ns* orbital together with the *np* orbitals and says that quadratic XeF_4 has two lone-pairs. For an experimentalist, this is not worth a dispute, but from a "ligand field" point of view, the diamagnetic quadratic p^2 systems Br(III), Te(II), I(III) and Xe(IV) and diamagnetic quadratic d^8 systems (some of the) Ni(II), Cu(III), Rh(I), Pd(II), Ag(III), Ir(I), Pt(II) and Au(III) have the point in common that they have a lone-pair perpendicular on the plane of the four ligands, with an angular function proportional to (z/r) or to $(3z^2/r^2) - 1$. As far goes lone-pairs, photo-electron spectra of gaseous molecules can provide interesting new information [103]. Everything seems fine in the case of one lone-pair. Thus, NH_3 has its lowest $I = 11.0$ and $N(CH_3)_3$ $I = 8.5$ eV, and similar results are known for many PX_3 and AsX_3. The lowest I of d-like orbitals in volatile compounds [64, 65, 102] are generally lower than these lone-pair values, but it is remembered that it is the opposite situation that creates a certain paradox [26, 102, 103, 107] for "ligand field" theory in iron(III) and copper(II) compounds, where anti-bonding d-like orbitals can have higher I than bonding M.O. of the same symmetry type. The higher I for 4f orbitals than of the highest filled M.O. is the rule rather than the exception in lanthanide compounds [26-28]. However, photo-electron spectra of gaseous molecules are conspicuously unable to detect two or more lone-pairs proposed by the Lewis paradigm. The water molecule has the lowest $I = 12.6$ eV corresponding to a non-bonding M.O. perpendicular on the molecular plane, being almost pure oxygen 2p in the L.C.A.O. model, and quite similar to the Xe5p lone-pair proposed above for XeF_4. However, the three following $I = 14.7$, 18.6 and 32.6 eV in H_2O are not at all good candidates for being a second lone-pair. The situation is perhaps slightly less conflictual in gaseous HX which should have three lone-pairs according to Lewis. The two π orbitals of HF having $I = 16.0$ eV and the σ with $I = 19.9$ eV are close to being F2p orbitals (and actually, the results of Price [99] can be considered [103] as a "ligand field" perturbation due to positive protons) whereas the σ orbital with $I = 39.6$ eV is closer to be a fluorine 2s orbital. It is unlikely that this orbital should be more important for bonding in monomeric HF than the 19.9 eV orbital, and it is a little ackward to say that this molecule has more than two lone-pairs.

A model based on early ideas of Lewis was proposed by Kimball, enclosing each pair of electrons in a sphere. The atom or molecule consist of spheres touching each other, and containing the nucleus in the smallest two-electron sphere of a given atom. Like in quantum mechanics, the kinetic energy of each electron confined in a sphere with radius R is proportional to R^{-2} and the proportionality constant is chosen by demanding the kinetic energy 1 rydberg $= \frac{1}{2}$hartree $=$ 13.6 eV for R $= \frac{3}{2}$bohr $= 0.79$ Å. This happens to be the average proton-electron distance in the groundstate of the hydrogen atom. This identification of an average value with the confining radius is not a good omen, and whereas the electronic density is supposed to be constant inside the Kimball sphere, the quantum-mechanical solution for the groundstate of one electron confined in a sphere is somewhat 1s-like, but the kinetic energy is $4\pi^2/9 = 4.38\ldots$ times larger than assumed in the Kimball model.[108]

The more pernicious side of the Kimball model[108] is that the total neglect of l-values allows a "Meccano construction" where the central small sphere is surrounded by a tetrahedron of four spheres (representing both the Lewis paradigm and the $K = 10$ of the neon atom) around which one may add six larger spheres, perhaps an allegoric representation of $N = 6$ in SF_6 and heavier elements. Since electrons anyhow are indiscernible fermions, their absolute confinement two and two looks artificial. It turns out[28, 108] that the inevitable coalescence of adjacent electron-pairs can be described by the "squared overlap integral" between two angular functions A_1 and A_2 which is obtained as the average value of $(A_1)^2(A_2)^2$ over a spherical surface (with the nucleus at the center). Two different angular functions belonging to the same l value (at least for p and d orbitals) have the "squared overlap integral" $(2l + 1)/(2l + 3)$, exactly a-third of the average value of $(A_1)^4$ for an angular function with itself. For comparison, it may be mentioned that the "squared overlap integral" between A_1 and a scalar s orbital is 1 since $(A_1)^2$ is normalized that way. These expressions should not be confused with the square of the overlap integral A_1A_2 which is zero, since the positive and negative contributions to A_1A_2 vanish on the spherical surface, when the two angular functions are orthogonal. It seems that the "squared overlap integral" gives a good representation of what a chemist means by two orbitals co-existing in our three-dimensional space, when centered on the same atom. The high values (3/5 for p and 5/7 for d orbitals) remove much of the confidence one might have had otherwise in the Kimball spheres as an heuristic model.

There has always been a tendency to approach the energy of a many-electron system (containing one or more nuclei) as a calculation of electrostatic potential energy modified by quantum conditions. However, the local operator of the kinetic energy[16, 109] is a much more relevant, and at the same time a much more specifically quantum-mechanical problem to consider for the chemist. One of the more unfamiliar aspects is that the kinetic energy depends implicitly on the shape of the orbitals, with the result that chemical bonding decreasing the average value of the second differential quotient with respect to all three Cartesian axes (as seen by bonding L.C.A.O.) is more influenced by considerations of kinetic than of potential energy. This is also the reason why the (tiny) non-spherical part of the Madelung potential broke down in 1956 as a plausible model[16, 87, 93] of "ligand field" effects in d-group compounds, and got replaced by the angular overlap model of partly covalent bonding.

8 Can Quantum Chemistry Save the Situation?

If need be, adult chemists can certainly survive without the Lewis paradigm. But it is legitimate to ask whether there may be some detail in quantum chemistry explaining why pairs of electrons are so appealing to chemists (in spite of the absence of significant attraction keeping the two electrons together). By the same token as we descended in deeper stratifications going from doubting the hybridization model to doubtling the Lewis paradigm, we may continue this excursion guided by Dante and look at a stratification below, common at least to M.O. theory and to the Lewis paradigm. The one-electron functions used as orbitals (also in Hartree-Fock functions of monatomic entities) accommodate two electrons with opposite spin-direction m_s in the non-relativistic Schrödinger equation (this is essentially also true for the relativistic treatment, but this would bring us far from our subject). An orbital occupied by both electrons feasible is a kind of electron-pair. However, a counter-acting influence is that higher symmetries [16] allow sets of two or three (in the two icosahedral point-groups even four or five) orbitals having the same energy of group-theoretical necessity (and sometimes, several orbitals have roughly the same energy "accidentally"). The extreme case is the full spherical symmetry. where the structure of the Periodic Table (however many complicated comments [10,18,21,28,33,110] one may add) is derived from $(2l + 1)$ orbitals having identical energy. If 2, 3, 4, ... (2k — 2) electrons occur in such a set of k degenerate orbitals, several energy levels lacking total symmetry occur. In the non-relativistic regime, well adapted to nearly all groundstates of chemical compounds, this means essentially to have $S = 1, 3/2, \ldots$ and not automatically 1/2 (for an odd number of electrons) nor zero. This evolution beyond the Lewis paradigm exhibits almost the atomic-spectroscopic aspect of J-levels of a partly filled 4f shell of lanthanide compounds [25–29]. It should be emphasized that high S and J values are not necessarily unstable alternatives, the extent of interelectronic repulsion [8,87] works in favor of high S values, like it does in monatomic entities involving partly filled shells (according to Hund's rules). It is tantalizing to the partisan of the Lewis paradigm that nuclei [15] have exactly this craving for total symmetry of the groundstate. Previously, they were said to contain Z protons and $(A - Z) = N$ neutrons (when the quarks [12–14] were invented in 1964, Z and N became more a kind of quantum numbers). With exception of three-quarter of the mass as individual protons, the large majority [111,112] of all nuclei in the Universe have simultaneously Z and N even. Such nuclei are all totally symmetric without positive spin.

Many text-books try to tame totally symmetric groundstates of many-electron systems by the remark that the one-electron wave-functions in an anti-symmetrized Slater determinant of a closed-shell configuration in a sense are arbitrary, because a unitary transformation of linear combinations of one-electron wave-functions (in the following: orbitals) does not modify the Slater determinant. By the way, this was the reason why penultimate M.O. frequently were considered calculational artifacts, until photo-electron spectra of gaseous molecules [97–99] became available. The whole argument has an Achilles heel: the total wave-function *is not* a Slater determinant. Though the groundstate of the helium atom [87] has a squared amplitude 0.99 of the configuration $1s^2$, this squared amplitude [21] varies between 0.9 and 0.8 for most atoms from beryllium to neon, and is probably below 0.5 for all atoms with Z higher than 30 (zinc). It is not easy to understand why this intermixing of configurations (due to non-diagonal elements of the interelectronic repulsion) does not prevent the existence of a functioning preponderant configuration [8]. One expects that these difficulties readily can be far worse in molecules than in atoms.

The writer is not an unconditional defender [113] of the Copenhagen interpretation of quantum mechanics, but it is certainly coherent to argue that the orbitals do not really manifest themselves before excitation or ionization modifying their full occupation. There was a great interest at one time to test the total electronic densities of monatomic Hartree-Fock wave-functions. This was mainly done by the one-parameter method of diamagnetism proportional to the average value of r^2 for the whole atom, but also by more detailed radial densities evaluated from electron diffraction. Anyhow, it would seem that the electronic density in our three-dimensional space is well represented. This is much less true for the interelectronic repulsion energy differences in a partly filled d-shell [8,87,115] which are typically $(z + 2)/(z + 3)$ times the Hartree-Fock results calculated for a gaseous ion with charge $+z$. The discrepancy is slightly worse in the 4f group. The only consolation is that this decrease from integrals to parameters of the Slater-Condon-Shortley theory (later stream-lined by Racah) gives the impression of a dielectric effect operating in the six-dimensional space needed to describe the simultaneous position of two electrons. One corollary of this fact is that chemists cannot obtain fulfilment of the wish that all energies are determined by electron densities in a three-dimensional space alone. Quantum mechanics is much more extravagant, prescribing 3 spatial variables and one spin variable for each electron. However, since we do not seem to need operators involving more than two electrons at the time, we may possibly construct a 6-dimensional simplification, though the appropriate criterium corresponding to the variational principle seems tough to find. The narrow-line luminescence of Gd(III) at 4 eV in the ultra-violet [27] and in the red at 1.6 eV (= 37 kcal/mole) of ruby $Al_{2-x}Cr_xO_3$ is almost exclusively due to lower interelectronic repulsion in the groundstate than in the excited state, and the electronic densities are very close to identical. A further surprise to chemists is that the excited state has S one unit lower than the groundstate, having $S = 7/2$ and $3/2$ respectively.

There have been many attempts to introduce geminals [114], two-electron functions with some built-in correlation and dielectric effects, as starting materials (replacing orbitals) for many-electron wave-functions. It is very difficult to ascertain the success of such an enterprise. At one hand, it cannot be avoided that a lot of free parameters ameliorate an Ansatz. On the other hand, according to Gombás and Gaspar. the total energy of atoms with Z between 5 and 90 is quite close to $-Z^{2.40}$ rydberg (it is not likely that the exponent is exactly 12/5 because its asymptotic value for very large Z is 7/3 in the Thomas-Fermi model, which is somewhat unrealistic in this limit by the non-relativistic requirement of vanishing $1/c$) and it turns out [16] that the correlation energy (relative to the Hartree-Fock function) is proportional to the square-root of this expression, $-Z^{1.2}$ with a proportionality constant of some 0.7 eV meaning that the correlation energy is roughly 40 eV for zinc and 130 eV for mercury. It is clear that a geminal treatment is attractive if it can recover half of this correlation energy. Since typical bond dissociation energies are 3 to 6 eV, it is of interest to note that the typical intra-atomic relaxation energy for I of inner shells [101] (i.e. the decrease of I relative to the Hartree-Fock value, due to the Manne-Åberg principle that photo-electron spectra indicate the eigen-values of the *new* Hamiltonian after the ejected electron already has left) is 0.8 eV times the square-root of I (in eV) and typical inter-atomic relaxation energies are 5 eV. This is the order of magnitude of chemical shifts of inner-shell I of a given element [100] to which inter-atomic relaxation gives a contribution of the same size (but frequently of the opposite sign) as the effects of varying fractional atomic charge and of differing Madelung potential [102]. A pure example of inter-atomic relaxation is the decrease of all inner-shell I values by 3 eV when mercury atoms are condensed to the metal.

An approach almost going back to the earliest ideas of Lewis was introduced by Linnett that the two manifolds of valence electrons with opposite m_s, have a tendency to form two regular tetrahedra, locked together in a typical Lewis molecule like methane, but moving more freely relative to each other in molecules with positive S (such as NO, NO_2 and O_2) and in the neon atom. However, a closer analysis [115] shows that one has to be very careful not to confuse various average values with the most frequent value. There is absolutely nothing tetrahedral about the spherical groundstate of the neon atom, though it is true [8,21] that the Hartree-Fock wave-function has less interelectronic repulsion than the simple Hartree product function without anti-symmetrization, and the configuration mixture even less, corresponding approximately to the correlation energy.

Perhaps the most constructive contribution to solve this problem is the study of pair-densities in hydrides of elements from lithium to fluorine by Bader and Stephens [116]. One major advantage is that it is not a predetermined working hypothesis, but a quantitative test of validity of Lewis structures, which may be appropriate or not. Bader and Stephens conclude that spatially localized pairs are satisfactory in LiH, BeH_2, BH_3 and BH_4^-; that CH_4 (among all molecules) is a border-line case, and that intra-correlated pair-functions would fail to recover a major fraction of the correlation energy in NH_3, H_2O and HF. It is also true for the neon atom and for N_2 and F_2 that most of the correlation energy comes from correlation between even the best optimized electron-pairs. There is no physical basis for the view that there are two separately localized pairs of non-bonded electrons in H_2O.

Bader, Beddall and Cade [117] discussed partitioning and characterization of molecular charge distributions. They suggested to consider the point along the internuclear axis at which the electronic density attains its minimum value between a pair of bonded nuclei, as the signpost indicating the border between two atoms. This may have an "inverted Robin Hood" effect, taking from the poor and giving to the rich, in so far the outer part of an atomic core with many electrons tend to push this minimum farther away from its own nucleus. These considerations (remaining in a three-dimensional space) have been further developed in a treatment of quantum topology in B_2H_6 and polyatomic hydrocarbon molecules [118-121].

Bader and Nguyen-Dang [122] have also written an interesting re-assessment of the Dalton idea of atoms in molecules, when going to recent quantum chemistry. These authors regret that the modern consensus is to regard atoms and bonds, and thus the structure itself, as useful but undefinable concepts. Disregarding the facts that most inorganic compounds in the condensed states do not consist of molecules, and that many interesting polyatomic species are cations or anions, it cannot be disputed away that the Z-values of the nuclei are the fundamental carriers of chemical characteristics. A secondary invariant carrier are the atomic cores present from lithium on, but in a way indefinite in hydrogen atoms. However, the writer would like to add that the density of uncompensated spin is a legitimate quantum-mechanical observable in systems with positive S, and that both nitrogen- and oxygen-containing molecules, and d- and f-group compounds, with positive S generally have only one set of M.O. partly filled, with the result that the spin-density becomes a picture of the orbital. This statement has a few exceptions in diatomic species such [123] as C_2^+ with the groundstate having $S = 3/2$ and [124] in TiO with $S = 1$, where two sets of M.O. are carriers of uncompensated spin. Whereas $3d^5$ Mn(II) and Fe(III) compounds with $S = 5/2$ are closely related to monoatomic entities in spherical symmetry, diatomic molecules [125] such as MnH and MnF with $S = 3$ are more similar to the lowest configuration $3d^54s$ of gaseous Mn^+. But even rare exceptions may sometimes tell us something interesting about chemical bonding.

9 Kossel versus Lewis, Present Status

The Lewis paradigm has two sides, the sum of the number of lone-pairs and of the number N of nearest-neighbor atoms normally being 4; and chemical bonds being effectuated by 2 electrons. Among inorganic compounds, only a-third or so have N at most 4. This is an observation relatively close to objective facts, however much many cases have an element with less well-defined N due to a wide dispersion of internuclear distances R, and due to legitimate discrepancies between time-average crystal structures and various spectroscopic techniques with differing short time-scales. There is no sharp limit for the validity, as Z increases, BiI_4^- has still $N = 4$ though Gillespie tells us that there is an additional lone-pair. On the other hand, the troubles start very early, as seen in NaCl-type carbides, nitrides and oxides with $N = 6$.

The question of two-electron bonds is trivial in H_2. But short hydrogen bonds and bent hydride bridges have $N = 2$, and many solid non-metallic hydrides $N = 6$. There has been so much effort spent the last half century to find the glue between two electrons inside a pair, and explain their propensity to form cylindrical zeppelins (if not cigarettes) between two adjacent atoms (because it is a condition in the job description for becoming chairman for the Department of Chemistry in El Dorado) that we now know a lot about why it does not work. Since there are probably more inorganic compounds having $N = 6$ than 4, the flexibility of M.O. theory of admitting highly different delocalization in L.C.A.O. of differing symmetry type has removed much of the motivation behind the two-electron bonds.

The approach of Kossel (also starting in 1916) has been far more fruitful for understanding the chemistry of 90 elements (admitting that about 10 other elements form many more organic than inorganic compounds). The repetitive pattern in the Periodic Table is due to closed shells having the total symmetry type which is the neutral element of Hund vector-coupling [22,23] quite independent of the exact nature of the highly complicated many-electron wave-function. Hence, the relations with atomic spectra are very important, and also the ionization energies of gaseous ions, once it was accepted [8,24,33,74] that the removal of an electron from a species in aqueous solution to vacuum needs the standard oxidation potential E^0 to which is added 4.5 eV (known with a precision of a-tenth eV). However, in hindsight, the most innovating endeavor of Kossel [20] were the connections with inner shells, which he studied with X-ray spectra, and which are even more informative to study with photo-electron spectra [21,100–104]. The number K of electrons in a Kossel isoelectronic series used to define the oxidation state $z = (Z - K)$ is not meant to indicate electrovalent bonding in an exclusive sense (it is known [8] from absorption spectra of d-group complexes that most central atoms carry fractional charges in the interval $+1$ to $+2$) but describes the symmetry types of the groundstate and of the lower excited states in the form of a preponderant electron configuration. Whereas nearly all f-group and non-metallic d-group compounds in this sense have a definite number of electrons in the partly filled shell, there remains the part of truth in the rule of Abegg that s- and p-shells are more mixed up by chemical bonding, and that the distinction between l values required in spherical systems no longer is operating, though it remains to a certain extent in photo-electron spectra [103].

It is a sociological fact that since 1916, Lewis has a far wider audience in textbooks than Kossel. This has several origins. In spite of the far larger number of organic compounds, they were much earlier classified using tetrahedral $N = 4$ (and explaining lower N by multiple bonds, etc.) and systematized by functional groups. This seemed hopeless for the fewer, but much more individualized, inorganic compounds. It should be remembered that Werner started with nitrogen stereochemistry and then went to octahedral $N = 6$ for cobalt(III) and many other central atoms. Lewis had the luck to occupy the arena just before crystal structures appeared. A professional crystallographer, Pauling, chose to elaborate the Lewis paradigm into the hybridization model, which became the favored tool for magnetochemists (who were the theoretical avant-garde of inorganic chemistry 1932—1954). This epoch finished when visible spectra came into focus. There may also be a psychological reason. Kossel did not want to break off the relations with atomic spectroscopy, and may have been considered unnecessarily complicated. It is a stress for many chemists not to possess a fixed set of axioms and a rounded-off fundament for everything, and they are willing to accept an imitation lasting for their life-time. After all, many people prefer Khomeiny rather than Hamlet.

There are not many analogous comprehensive postulates in chemistry, and they are in a way grandiose, even when they fail. The closest analogy to the Lewis paradigm is probably the idea of Lavoisier that all acids contain oxygen. This is still true for about half of the acids, but it is no longer considered pertinent.

10 References

1. Lewis, G. N.: J. Am. Chem. Soc. *38*, 762 (1916)
2. Jørgensen, C. K.: Chimia *25*, 109 (1971)
3. Jørgensen, C. K.: Chimia *28*, 605 (1974)
4. Jørgensen, C. K.: Revue chim. min. (Paris) *20*, 533 (1983)
5. Jørgensen, C. K.: Chimia *38*, 75 (1984)
6. Daumas, M.: Lavoisier, théoricien et expérimentateur. Paris: Presses Universitaires de France 1955
7. Weeks, M. E.: The Discovery of the Elements (7. ed.). Easton, Penn.: J. Chem. Educ. Publ. 1968
8. Jørgensen, C. K.: Oxidation Numbers and Oxidation States. Springer: Berlin, Heidelberg, New York 1969
9. Palmer, W. G.: A History of the Concept of Valency to 1930. Cambridge: University Press 1965

10. Jørgensen, C. K.: J. chim. physique *76*, 630 (1979)
11. Pais, A.: Rev. Mod. Phys. *49*, 925 (1977)
12. Jørgensen, C. K.: Naturwissenschaften *69*, 420 (1982)
13. Jørgensen, C. K.: Nature *305*, 787 (1983)
14. Jørgensen, C. K.: Naturwissenschaften *71*, 151 (1984)
15. Jørgensen, C. K.: Structure and Bonding *43*, 1 (1981)
16. Jørgensen, C. K.: Modern Aspects of Ligand Field Theory. Amsterdam: North-Holland 1971
17. Van Spronsen, J. W.: The Periodic System of Chemical Elements. Amsterdam: Elsevier 1969
18. Jørgensen, C. K.: Radiochim. Acta *32*, 1 (1983)
19. Dasent, W. E.: Nonexistent Compounds. New York: Dekker 1965
20. Kossel, W.: Ann. Physik *49*, 229 (1916)
21. Jørgensen, C. K.: Adv. Quantum Chem. *11*, 51 (1978)
22. Jørgensen, C. K.: Israel J. Chem. *19*, 174 (1980)
23. Jørgensen, C. K.: Int. Rev. Phys. Chem. *1*, 225 (1981)
24. Jørgensen, C. K.: Comments Inorg. Chem. *1*, 123 (1981)
25. Jørgensen, C. K.: Gmelin Handbuch der anorganischen Chemie: Seltenerdelemente *B1*, 17 (Springer-Verlag: 1976)
26. Jørgensen, C. K.: Structure and Bonding *22*, 49 (1975)
27. Reisfeld, R., Jørgensen, C. K.: Lasers and Excited States of Rare Earths. Springer: Berlin, Heidelberg, New York 1977
28. Jørgensen, C. K.: Handbook on the Physics and Chemistry of Rare Earths (eds. K. A. Gschneidner and L. Eyring) *3*, 111, Amsterdam: North-Holland 1979
29. Jørgensen, C. K., Reisfeld, R.: Topics Current Chem. *100*, 127 (1982)
30. Jørgensen, C. K., Reisfeld, R.: Structure and Bonding *50*, 121 (1982)
31. Jørgensen, C. K., Reisfeld, R.: J. Electrochem. Soc. *130*, 681 (1983)
32. Jørgensen, C. K.: Chem. Phys. Letters *2*, 549 (1968)
33. Jørgensen, C. K.: Chimia *23*, 292 (1969)
34. Johnson, D. A.: Adv. Inorg. Chem. Radiochem. *20*, 1 (1977)
35. Kauffman, G. B.: J. Chem. Educ. *36*, 521 (1959)
36. Werner, A.: Neuere Anschauungen auf dem Gebiete der anorganischen Chemie (3. Auflage). Braunschweig: F. Vieweg 1913
37. Kauffman, G. B.: Alfred Werner, Founder of Co-ordination Chemistry. Springer: Berlin, Heidelberg, New York 1966
38. Jørgensen, C. K.: Inorganic Complexes. London: Academic Press 1963
39. Jørgensen, C. K.: Acta Chem. Scand. *9*, 1362 (1955); *10*, 887 (1956)
40. Bjerrum, J.: Metal Ammine Formation in Aqueous Solution. Copenhagen: P. Haase 1941 (2. ed. 1957)
41. Bjerrum, J.: Chem. Rev. *46*, 381 (1950)
42. Rasmussen, L., Jørgensen, C. K.: Acta Chem. Scand. *22*, 2313 (1968)
43. Parthasarathy, V., Jørgensen, C. K.: Chimia *29*, 210 (1975)
44. Texter, J., Hastreiter, J. J., Hall, J. L.: J. Phys. Chem. *87*, 4690 (1983)
45. Gans, P., Gill, J. B.: J.C.S. Dalton 779 (1976)
46. Day, P., Hall, I. D.: J. Chem. Soc. (A) 2679 (1970)
47. Truter, M. R.: Annual Reports *76C* (for 1979) 161. London: Royal Society of Chemistry 1980
48. Tables of Interatomic Distances and Configuration in Molecules and Ions, Special Publications no. 11 and 18. London: Chemical Society 1958 and 1965
49. Murray-Rust, P., Bürgi, H. B., Dunitz, J. D.: J. Am. Chem. Soc. *97*, 921 (1975)
50. Bye, E., Schweizer, W. B., Dunitz, J. D.: J. Am. Chem. Soc. *104*, 5893 (1982)
51. Schwarzenbach, D.: Chimia *37*, 373 (1983)
52. Pauling, L.: J. Am. Chem. Soc. *55*, 1895 (1933)
53. Jørgensen, C. K., Rittershaus, E.: Mat. fys. Medd Danske Vid. Selskab (Copenhagen) *35*, no. 15 (1967)
54. Sinha, S. P.: Structure and Bonding *25*, 69 (1976)
55. Hoffmann, R., Beier, B. F., Muetterties, E. L., Rossi, A. R.: Inorg. Chem. *16*, 511 (1977)
56. Favas, M. C., Kepert, D. L.: Progress Inorg. Chem. *28*, 309 (1981)
57. Magnus, A.: Z. anorg. Chem. *124*, 289 (1922); Physik. Z. *23*, 241 (1922)
58. Martin, T. P.: Phys. Reports *95*, 167 (1983)

59. Barton, L.: Topics Current Chem. *100*, 169 (1982)
60. Malatesta, L.: Progress Inorg. Chem. *1*, 283 (1959)
61. Jørgensen, C. K.: Inorg. Chem. *3*, 1201 (1964)
62. Willemse, J., Cras, J. A., Steggerda, J. J., Keijzers, P.: Structure and Bonding *28*, 83 (1976)
63. Gmelin Handbuch der anorganischen Chemie: Ergänzungswerk zur 8. Auflage, vol. 3: Chromorganische Verbindungen (Springer-Verlag 1971)
64. Furlani, C., Cauletti, C.: Structure and Bonding *35*, 119 (1978)
65. Green, J. C.: Structure and Bonding *43*, 37 (1981)
66. Warren, K. D.: Structure and Bonding *27*, 45 (1976)
67. Marks, T. J.: Science *217*, 989 (1982)
68. Jørgensen, C. K.: Chem. Phys. Letters *87*, 320 (1982)
69. Tachikawa, M., Muetterties, E. L.: Progress Inorg. Chem. *28*, 203 (1981)
70. Giguère, P. A.: J. Chem. Educ. *56*, 571 (1979)
71. Giguère, P. A.: Revue chim. min. (Paris) *20*, 588 (1983)
72. Giguère, P. A.: Chem. Phys. Letters *41*, 598 (1976)
73. Jørgensen, C. K.: Naturwissenschaften *67*, 188 (1980)
74. Jørgensen, C. K.: Topics Current Chem. *56*, 1 (1975)
75. Latimer, W. M.: J. Chem. Phys. *23*, 90 (1955)
76. Morris, D. F. C.: Structure and Bonding *4*, 63 (1968); *6*, 157 (1969)
77. Beutler, P., Gamsjäger, H., Baertschi, P.: Chimia *32*, 163 (1978)
78. Kauffman, G. B., Baxter, J. F.: J. Chem. Educ. *58*, 349 (1981)
79. Hunt, J. P., Friedman, H. L.: Progress Inorg. Chem. *30*, 359 (1983)
80. Romano, V., Bjerrum, J.: Acta Chem. Scand. *24*, 1551 (1970)
81. Jørgensen, C. K., Parthasarathy, V.: Acta Chem. Scand. *A 32*, 957 (1978)
82. Hewish, N. A., Neilson, G. W., Enderby, J. E.: Nature *297*, 138 (1982)
83. Narten, A. H., Hahn, R. L.: J. Phys. Chem. *87*, 3193 (1983)
84. Skibsted, L. H., Bjerrum, J.: Acta Chem. Scand. *A 28*, 740 and 764 (1974)
85. Penneman, R. A., Mann, J. B., Jørgensen, C. K.: Chem. Phys. Letters *8*, 321 (1971)
86. Penneman, R. A., Mann, J. B.: Proceed. Moscow Symposium on the Chemistry of Transition Elements 1972 (eds. V. I. Spitsyn and J. J. Katz), Suppl. J. Inorg. Nucl. Chem. p. 257, Oxford: Pergamon 1976
87. Jørgensen, C. K.: Orbitals in Atoms and Molecules. London: Academic Press 1962
88. Lackner, K. S., Zweig, G.: Phys. Rev. *D 28*, 1671 (1983)
89. Jørgensen, C. K.: Structure and Bonding *34*, 19 (1978)
90. Jørgensen, C. K.: J. Phys. Chem., submitted
91. De Rújula, A., Giles, R. C., Jaffe, R. L.: Phys. Rev. *D 17*, 285 (1978)
92. Williams, A. F.: Theoretical Approach to Inorganic Chemistry. Springer: Berlin, Heidelberg, New York 1979
93. Jørgensen, C. K.: Absorption Spectra and Chemical Bonding in Complexes. Oxford: Pergamon 1962 (2. ed. 1964)
94. Jørgensen, C. K.: Adv. Chem. Phys. *5*, 33 (1963)
95. Jørgensen, C. K.: Progress Inorg. Chem. *12*, 101 (1970)
96. Herzberg, G.: Electronic Spectra and Electronic Structure of Polyatomic Molecules. New York: Van Nostrand 1966
97. Turner, D. W., Baker, C., Baker, A. D., Brundle, C. R.: Molecular Photoelectron Spectroscopy. London: Wiley-Interscience 1970
98. Rabalais, J. W.: Principles of Ultraviolet Photoelectron Spectroscopy. New York: Wiley-Interscience 1977
99. Brundle, C. R., Baker, A. D. (eds.): Electron Spectroscopy (vol. 1). London: Academic Press 1977
100. Jørgensen, C. K., Berthou, H.: Mat. fys. Medd. Danske Vid. Selskab (Copenhagen) *38*, no. 15 (1972)
101. Jørgensen, C. K.: Adv. Quantum Chem. *8*, 137 (1974)
102. Jørgensen, C. K.: Structure and Bonding *24*, 1 (1975)
103. Jørgensen, C. K.: Structure and Bonding *30*, 141 (1976)
104. Jørgensen, C. K.: Fresenius Z. analyt. Chem. *288*, 161 (1977)
105. Jørgensen, C. K.: Chem. Phys. Letters *27*, 305 (1974)
106. Gillespie, R. J.: Molecular Geometry. New York: Van Nostrand-Reinhold 1972

107. Jørgensen, C. K.: Chimia *27*, 203 (1973); *28*, 6 (1974)
108. Jørgensen, C. K.: Chimia *31*, 445 (1977)
109. Ruedenberg, K.: Rev. Mod. Phys. *34*, 326 (1962)
110. Katriel, J., Jørgensen, C. K.: Chem. Phys. Letters *87*, 315 (1982)
111. Trimble, V.: Rev. Mod. Phys. *47*, 877 (1975)
112. Kuroda, P.: The Origin of the Chemical Elements. Springer: Berlin, Heidelberg, New York 1982
113. Jørgensen, C. K.: Theor. Chim. Acta *34*, 189 (1974)
114. Levy, M.: J. Chem. Phys. *61*, 1857 (1974)
115. Jørgensen, C. K.: Solid State Phys. *13*, 375 (1962)
116. Bader, R. F. W., Stephens, M. E.: J. Am. Chem. Soc. *97*, 7391 (1975)
117. Bader, R. F. W., Beddall, P. M., Cade, P. E.: J. Am. Chem. Soc. *93*, 3095 (1971)
118. Bader, R. F. W., Anderson, S. G., Duke, A. J.: J. Am. Chem. Soc. *101*, 1389 (1979)
119. Bader, R. F. W., Tal, Y., Anderson, S. G., Nguyen-Dang, T. T.: Israel J. Chem. *19*, 8 (1980)
120. Bader, R. F. W.: J. Chem. Phys. *73*, 2871 (1980)
121. Bader, R. F. W., Ting-Hua Tang, Tal, Y., Biegler-König, F. W.: J. Am. Chem. Soc. *104*, 940 and 946 (1982)
122. Bader, R. F. W., Nguyen-Dang, T. T.: Adv. Quantum Chem. *14*, 63 (1981)
123. Verhaegen, C.: J. Chem. Phys. *49*, 4696 (1968)
124. Jørgensen, C. K.: Mol. Phys. *7*, 417 (1964)
125. Van Zee, R. J., De Vore, T. C., Wilkerson, J. L., Weltner, W.: J. Chem. Phys. *69*, 1869 (1978)

Cationic and Anionic Complexes of the Noble Gases

Henry Selig[1] and John H. Holloway[2]

1 Department of Inorganic and Analytical Chemistry, The Hebrew University of Jerusalem, Jerusalem, Israel
2 Department if Chemistry, The University of Leicester, Leicester LE 1 7RH, England

Table of Contents

A Introduction

The existence of simple compounds of the noble gases was first demonstrated in 1962 [1]. Although it was soon recognized that the binary xenon fluorides, particularly XeF_2 and XeF_6, could form numerous adducts with fluoride-ion acceptors and, in some cases, with donors, the true nature of the complexes was not fully appreciated at first. Stimulated by the initial surprise at the existence of the compounds, early work was devoted rather to evaluation of the nature of the simple compounds and investigation of their physical properties.

It was quickly realized that the fluorides and oxide fluorides of the noble gases bear remarkable similarities in their chemical and structural properties to those of the neighbouring halogen compounds. As various spectroscopic techniques such as X-ray crystallography, vibrational spectroscopy and multinuclear magnetic resonance began to be applied, the nature of the complexes was elucidated and analogies to the halogen compounds confirmed.

Although there has been controversy as to the extent of participation of outer noble-gas orbitals in bonding, the valence shell electron pair repulsion theory, which invokes such involvement, has proved to be a useful tool for predicting the structures of the binary compounds, the oxide fluorides, and cationic and anionic derivatives where they exist. For compounds with coordination number of six or less (including non-bonding electron pairs) this model predicts, without fail, the correct geometry. Only for coordination numbers exceeding six and for the larger ligands does the model begin to break down, as can be seen for the species XeF_6, $[XeF_8]^{2-}$ and $[TeCl_6]^{2-}$. Descriptions of structures of different xenon species according to coordination number, number of ligands and number of nonbonding electron pairs have been tabulated elsewhere [2].

At present most of the known chemistry of noble gases involves formation of various types of adducts, but a systematic description of this chemistry has been lacking because most of this knowledge, particularly for the KrF_2 adducts, has been developed only during the last decade.

Known complexes can be classified as derivatives of binary fluorides or oxide fluorides according to the following scheme (see next page).

Additionally, there are a number of somewhat more "exotic" species such as, $[Xe_2]^+$, $[XeOTeF_5]^+$, $[(XeO)_2S(O)F]^+$, $([XeN(FO_2S)_2]_2F)^+$ and $[XeO_3X]^-$ (X = F, Cl, Br).

The formulation of complexes as salt-like species containing well-defined cations or anions should be approached with a certain amount of caution. This is particularly well illustrated with xenon difluoride adducts which span the gamut of complexes from salt-like species such as $[XeF]^+[Sb_2F_{11}]^-$ to covalent adducts like $XeF_2 \cdot XeOF_4$. In the latter the components preserve their molecular identities and dimensions and the adduct is clearly a covalent adduct, but even in the former the relative short Xe $\cdots$ F distance between the $[XeF]^+$ cation and the $[Sb_2F_{11}]^-$ anion (2.34 Å) implies considerable covalent character.

Criteria which have been drawn upon in classifying the compounds are based mainly on bond distances obtained from X-ray crystal structure determinations, from the stretching frequencies of both terminal and bridging fluorine bonds, and

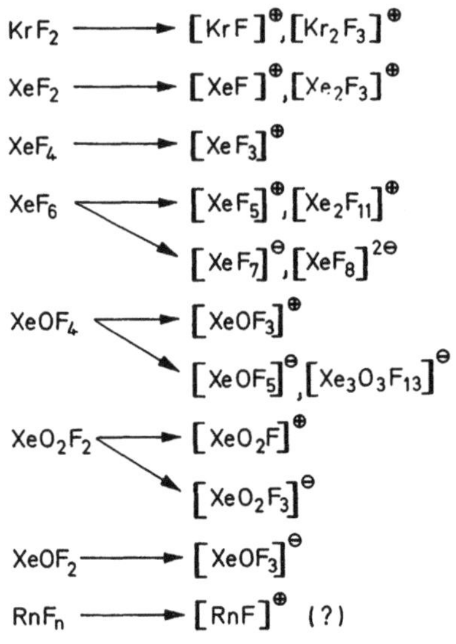

from ^{19}F and [129]Xe n.m.r. chemical shift data. Clear correlations exist between all of these.

The compounds XeF_2 and XeF_6 form numerous stable complexes with fluoride-ion acceptors having compositions, $2\,XeF_n \cdot MF_5$, $XeF_n \cdot MF_5$ and $XeF_n \cdot 2\,MF_5$ (n = 2 or 6). The corresponding chemistry of XeF_4 is considerably more sparse and, as a result, it has been concluded that XeF_4 is a much weaker fluoride-ion donor and that fluoride ion donor ability decreases in the sequence $XeF_6 > XeF_2 \gg XeF_4$ [3]. However, the relative order of fluoride ion donor ability of XeF_2 and XeF_4 has been disputed. Using as criteria a) the average length of the Xe ··· F bridge, which decreases in the series $[F_5Xe]^+ \cdots F$, $[F_3Xe]^+ \cdots F$ and $[FXe]^+ \cdots F$, b) the direction of the bridge bonds, which form in directions which avoid both the bonding pairs and the presumed location of non-bonding pairs in the valence shell, and c) the number of bridge bonds, which increase with decreasing strength of the bridge bonds, the fluoride ion acceptor abilities of the ions have been graded as $XeF_6 > XeF_4 > XeF_2$ [4].

B Cations

I $[Xe_2]^+$

A transient green colour in solutions of XeF_2—XeF_4—SbF_5 [5] reported many years ago has recently been identified with the dixenon (+1) cation, $[Xe_2]^+$ [6,7]. This ion had been observed previously only in gaseous xenon as the product of excited atom collisions. The dixenon cation is a product of either oxidation of gaseous xenon by the dioxygenyl species or by reduction of $[XeF]^+$ (+2) with water or other reducing agents according to the scheme:

$$2\,Xe \xrightarrow[{[O_2]^{\oplus}}]{-e} [Xe_2]^{\oplus} \xleftarrow[{H_2O \text{ or Pb or Hg}}]{+e} 2\,[XeF]^{\oplus}$$

Solutions of the green species in SbF_5 are indefinitely stable at ambient temperature so long as they are maintained under a pressure of xenon [6–8].

Formation of the species has been monitored by e.s.r., u.v.-visible and Raman spectroscopy. A simulated e.s.r. spectrum best fits that observed in an SbF_5 matrix at 4.5 K for the values $g_{\perp} = 2.304$, $g_{\parallel} = 1.885$. The u.v.-visible spectrum of $[Xe_2]^+$ shows, in addition to the $[XeF]^+$ absorption at 287 nm, two absorptions at 335 and 720 nm [7], the optical densities of which are proportional to the $^3/_2$ power of both the xenon pressure and the $[XeF]^+[Sb_2F_{11}]^-$ concentration [7,8]. This has been interpreted in terms of the following simultaneous reactions (Eqs. 1 and 2):

$$3\,Xe + [XeF]^+[Sb_2F_{11}]^- + 2\,SbF_5 \rightleftharpoons 2\,[Xe_2]^+[Sb_2F_{11}]^- \quad (1)$$

$$[Xe_2]^+[Sb_2F_{11}]^- + [XeF]^+[Sb_2F_{11}]^- \rightleftharpoons [Xe_2]^+[Sb_2F_{11}]^- \cdot [XeF]^+[Sb_2F_{11}]^- \quad (2)$$

that is, formation and complexation of the dication complex with excess of oxidant. The overall reaction is thus (Eq. 3):

$$3\,Xe + 3\,([XeF]^+[Sb_2F_{11}]^-) + 2\,SbF_5 \rightleftharpoons 2\,([Xe_2]^+[Sb_2F_{11}]^- \cdot [XeF]^+[Sb_2F_{11}]^-) \quad (3)$$

The dicationic species can be oxidized by fluorine (Eq. 4):

$$[Xe_2]^+[Sb_2F_{11}]^- \cdot [XeF]^+[Sb_2F_{11}]^- + {}^3/_2\,F_2 + 2\,SbF_5 \rightarrow 3\,[XeF]^+[Sb_2F_{11}]^- \quad (4)$$

These reactions can be reversed by either adding an excess of xenon or an excess of fluorine, respectively. In the latter case the yellow $[XeF]^+[Sb_2F_{11}]^-$ solutions are obtained.

Formation of the dixenon cation can also be monitored by the appearance of a band at 123 cm^{-1} in the Raman spectrum. A striking resemblance to the dark green isoelectronic $[I_2]^-$ is to be noted with strong absorptions at 370–400 and 737–800 nm in the visible and $w_e = 114$–$116\ cm^{-1}$ in the resonance Raman spectrum of $M^+[I_2]^-$ [9].

II $[ArF]^+$, $[KrF]^+$, $[XeF]^+$, $[RnF]^+$, $[Kr_2F_3]^+$ and $[Xe_2F_3]^+$

1 Introduction

Divalent noble-gas compounds are undoubtedly the most intensively studied species in noble-gas chemistry and the divalent fluorides, NgF_2 (Ng = Kr, Xe or Rn), is the only class common to all three elements. Although radon is expected to form compounds more readily than xenon, it is available only as an intensely radioactive isotope (^{222}Rn, half-life 3.83 days) and, consequently, rather little has been done with it. As far as krypton and xenon are concerned the molecular orbital ordering of KrF_2 is rather similar to that of XeF_2 [10] but photoionisation mass spectrometric studies and calorimetric data concur that the average bond energy in KrF_2 is much smaller (Table 1). The +2 oxidation state for krypton and xenon is more stabilized in the diatomic noble-gas fluoride cations and here photoionisation mass spectrometric data have shown that D_0 ($[KrF]^+$) $\leqq$ 1.63 eV [12] compared with D_0 ($[XeF]^+$) = 2.03 eV [14]. This clearly shows that the average bond energy of the diatomic

Table 1. Comparison of Photoionization Mass Spectrometric and Calorimetric Data on KrF_2 and XeF_2

	Mean thermochemical bond energy	Average bond energy from photoionization mass spectrometry
KrF_2	49.0 KJ mol^{-1} [a]	44.4 KJ mol^{-1} [b]
XeF_2	133.9 KJ mol^{-1} [c]	135.4 KJ mol^{-1} [d]

[a] Ref. [11] [b] Ref. [12] [c] Ref. [13] [d] Ref. [14]

molecular ion is significantly greater than that of the triatomic neutral species and prompted a search for values for the series of $[NgF]^+$ molecular ions as a means of assessing the possibilities of synthesizing them. In these experiments $[HeF]^+$ and $[NeF]^+$ were not observed but $[ArF]^+$ was produced and led to a value of D_0 ($[ArF]^+$) $\geqq$ 1.655 eV [15]. Argon compounds have still not been successfully isolated but the production of $[ArF]^+$-containing adducts with counter-anions derived from highly electronegative molecules such as PtF_5 and AuF_5 presents the best experimental possibility for the preparation of an argon species.

Although compounds of composition $KrF_2 \cdot 2\,SbF_5$ [16] and $XeF_2 \cdot 2\,SbF_5$ [17] have been known since almost the beginning of noble-gas chemistry, it has taken several years to understand the true nature of these and related compounds. It has now been shown that, in the solid state, the adducts $NgF_2 \cdot 2\,MF_5$, $NgF_2 \cdot MF_5$ and $2\,NgF_2 \cdot MF_5$ [(Ng = Kr with M = As, Sb, Bi, Nb, Ta or Au) [18–26]; (Ng = Xe with M = As, Sb, Bi, Nb, Ta, Ru, Os, Ir, Pt) [5,26–40]] can be thought of in terms of ionic formulations involving $[NgF]^+$ and $[Ng_2F_3]^+$ cations and $[M_2F_{11}]^-$ and $[MF_6]^-$ anions, but it is also clear that most of the adducts exhibit weak covalent interactions between the anion and the cation through fluorine bridging [26,33,36,39,40] and that the compounds should not, therefore, be regarded as simple salts. Indeed a crystal structure of what is probably the most ionic of the compounds, $XeF_2 \cdot 2\,SbF_5$ [30,31], has shown that the xenon to bridging-fluorine bond distance is 2.35 Å, which has been taken as evidence of sufficient bond formation to justify regarding the compound as essentially covalent. Structural studies on the related $XeF_2 \cdot MF_5$ [M = As, [39] Ru, [34]] and $2\,XeF_2 \cdot AsF_5$ [37] species invite similar interpretations since these, too, have rather short anion-cation contacts. No X-ray single-crystal data is available for the krypton-containing adducts, but the similarity of their vibrational spectra [22,23] to those of the xenon species [33,38] indicates close structural analogies.

A striking characteristic of the Raman spectra of the $XeF_2 \cdot MF_5$ and $XeF_2 \cdot 2\,MF_5$ melts (M = Sb, Ta, Nb) is the presistence of bands associated with the Xe ··· F and M ··· F bridging bonds. Indeed, the δ(F—Xe ··· F) bending modes are clearly visible in the spectra of most adducts [41]. Electrical conductivities of the same series of compounds, together with viscosity measurements on the adduct with the highest specific conductivity, $XeF_2 \cdot SbF_5$, have given a value of the order of 11% for the degree of dissociation of the compounds near the melting point [42]. This is of the same order as alkyl ammonium picrates, which are known to be incompletely dissociated; and so the melts also do not exhibit ideal salt behaviour.

Solution ^{19}F and ^{129}Xe n.m.r. studies have provided comprehensive data on the nature of both krypton and xenon species present [43–45]. In particular, for the xenon-containing species, empirical correlations among ^{129}Xe chemical shifts have provided a sensitive probe for assessing the degree of ionic character in the Xe—F bond [44,45]. In the specific case of $[XeF]^+$ cation plots of the ^{129}Xe chemical shift vs. the ^{19}F chemical shift of the terminal fluorine on xenon for a range of xenon (II) species containing fluorine bridges and oxygen bridges have shown that, in SbF_5 solution, $[XeF]^+$ approximates to being a free cation, being only very weakly fluorine bridged to the $[Sb_nF_{5n+1}]^-$ polyanion under these conditions. The data on other related species corroborate the postulated Xe—F bond ionicities and structural information from solid-state work [44,45].

When solid-state and solution studies are carried out on krypton and xenon difluoride adducts incorporating fluoride acceptors in the weak to intermediate range, such as WF_4O and MoF_4O, the results indicate that, in both phases, the adducts contain Ng ··· F ··· M (Ng = Kr or Xe) bridges and ionic contributions to the bonding are minimised [46–49].

The overall picture therefore is of series of adducts in which the extent of ionic character in terms of the presence of $[NgF]^+$ or $[Ng_2F_3]^+$ ions is related to the relative fluoride ion acceptor strengths of the complexing agents.

2 $[ArF]^+$

Diatomic noble gas ions and diatomic hydride ions involving argon have been observed since the 1930's [1] and $[ArN]^+$ [50] and $[ArI]^+$ [51] were observed in collision experiments in mass spectrometers in 1960. However, none of these species were isolable as stable solids. Following the discovery of stable krypton and xenon fluorides and, in particular, recognition of the enhanced stability of $[KrF]^+$ and $[XeF]^+$ in crystalline solids, there has been renewed interest in the possibility of obtaining other related species.

Ab initio calculations on $[HeF]^+$ and $[NeF]^+$ [52] showed that the ground states of these molecules are unstable but the results of similar studies on $[ArF]^+$ implied a stable ground state and it was suggested that the most likely compound to be synthesized would be $[ArF]^+[PtF_6]^-$ [53]. Additional experimental evidence for the stability of the $[HeF]^+$, $[NeF]^+$, $[ArF]^+$ series was sought using a collision technique. This produced evidence of $[ArF]^+$ and led to $D_0([ArF]^+) \geq 1.655$ e.V. but confirmed the instability of $[HeF]^+$ and $[NeF]^+$ in the electronic ground states [15]. Recent work involving a redefinition of electronegativity leading to corrections in the assignments of electronegativity values to the noble-gas elements have permitted comparison of the Allred and Rochow, Sanderson, Mulliken and Pauling electronegativity scales. The agreement between all four is striking and values (He, 5.2; Ne, 4.5; Ar, 3.2; Kr, 2.9; Xe, 2.4 and Rn, 2.1) show that the values for argon and krypton as rather close [54]. Thus continued efforts to synthesize a cationic argon fluoride complex still seem justified.

3 $[KrF]^+$ and $[Kr_2F_3]^+$

a Preparations and Structures

So far krypton has exhibited chemistry only for the +2 oxidation state and only krypton-fluorine compounds are known. All the known divalent krypton complexes consist of molecules which exhibit a significant amount of ionic character and can be described in terms of $[KrF]^+$ and $[Kr_2F_3]^+$ containing species. The more complex cation consists of two $[KrF]^+$ units joined by a fluorine bridge.

Although a range of xenon difluoride adducts, $XeF_2 \cdot 2\,MF_5$, $XeF_2 \cdot MF_5$ and $2\,XeF_2 \cdot MF_5$ were studied in the ten years after the discovery of noble-gas chemistry, difficulties in obtaining large samples of krypton difluoride and concern about the rapidity of decomposition of KrF_2 [55] resulted in only one related KrF_2 adduct, $KrF_2 \cdot 2\,SbF_5$, being reported [16]. Evidence of an unstable adduct with AsF_5 was also obtained [16]. In 1971 an extensive review on krypton difluoride chemistry referred to a number of krypton difluoride adducts but no details of preparations or characterizations were reported [56]. In the same year these authors claimed that $KrF_2 \cdot 2\,SbF_5$ is the only compound formed in the KrF_2—SbF_5—BrF_5 system [57]. However, since 1973, there has been considerable activity in the field and $[KrF]^+$ [18–26, 58–60] and $[Kr_2F_3]^+$ [18–20, 22, 23, 59] have been characterized both in the solid state [18–26, 59, 60] and in solution [18, 20, 22, 23, 59] and adducts containing them have been shown to possess powerful oxidising and fluorinating abilities [20, 21, 59]. All of the compounds fully characterized so far are adducts of KrF_2 with Lewis acid pentafluorides and all have the stoichiometries 2:1, 1:1, or 1:2. Complexes reported are $KrF_2 \cdot VF_5$ [61], $KrF_2 \cdot 2\,NbF_5$, $KrF_2 \cdot 2\,TaF_5$ [18,23], $KrF_2 \cdot 2\,SbF_5$ [16, 20, 22, 23, 57, 58], $KrF_2 \cdot TaF_5$ [18,23], $KrF_2 \cdot PtF_5$ [20,22], $KrF_2 \cdot AuF_5$ [21,25], $KrF_2 \cdot AsF_5$ [20, 22, 59], $KrF_2 \cdot SbF_5$ [19, 20, 22, 23, 60], $2\,KrF_2 \cdot TaF_5$ [23], $2\,KrF_2 \cdot AsF_5$ [20, 22, 59], $2\,KrF_2 \cdot SbF_5$ [19, 20, 22, 23, 59]. In addition, there are a number of adducts in which additional KrF_2 units appear to be weakly associated, probably with the cationic parts of the adducts. These are perhaps best formulated as $[xKrF_2 \cdot KrF]^+[Nb_2F_{11}]^-$ $[xKrF_2 \cdot KrF]^+ \cdot [Ta_2F_{11}]^-$ [23], $[xKrF_2 \cdot KrF]^+[SbF_6]^-$ [19, 20], $[xKrF_2 \cdot Kr_2F_3]^+[AsF_6]^-$ [20, 22] and $[xKrF_2 \cdot Kr_2F_3]^+[SbF_6]^-$ [22].

The adducts are prepared by a series of reactions commencing with the judicious combination of krypton difluoride with the appropriate pentafluoride either directly[22] or in bromine pentafluoride solvent [23]. In all cases the experimental conditions require careful control as outlined in the detailed reaction sequences published [22, 23]. In one case, $[KrF]^+[AsF_6]^-$, both α- and β-forms occur [20, 22]. The antimony pentafluoride adducts have also been prepared starting from $[BrF_4]^+[Sb_2F_{11}]^-$ [19, 23]. The Raman spectra of the adducts are well documented and the progress of reactions are best monitored by this means.

The only room temperature stable solid adducts are $KrF_2 \cdot 2\,SbF_5$ [16,22,23,58], $KrF_2 \cdot PtF_5$ [22] and $KrF_2 \cdot SbF_5$ [19,22,23], while the rest decompose. In some cases, for example $KrF_2 \cdot TaF_5$ and $2\,KrF_2 \cdot AsF_5$, decomposition takes place even at —60 °C [22,23]. For the Sb—Ta—Nb series, qualitative estimates of the trend in thermal stability under dynamic vacuum suggest the order $KrF_2 \cdot 2\,SbF_5 > KrF_2 \cdot SbF_5 > KrF_2 \cdot 2\,TaF_5 > 2\,KrF_2 \cdot SbF_5 > KrF_2 \cdot TaF_5 > KrF_2 \cdot 2\,NbF_5$. Considering the low thermal stability of $KrF_2 \cdot 2\,NbF_5$, which dissociates under dynamic vacuum even at —50 °C, and the above trend, it is not surprising that $KrF_2 \cdot NbF_5$ has not been obtained [23].

It is also noteworthy that some of the compounds which are stable at room temperature are unstable in solution. Thus, solutions of $KrF_2 \cdot PtF_5$ and $KrF_2 \cdot SbF_5$ in anhydrous HF, KrF_2 in SbF_5 and $2\,KrF_2 \cdot MF_5$ and $KrF_2 \cdot MF_5$ (M = As, Nb, Ta or Sb) in BrF_5 all decompose rapidly at room temperature liberating krypton and fluorine [22,23]. In the case of the BrF_5 solvent, some of the bromine pentafluoride is also oxidized to $[BrF_6]^+$ cation at room temperature [20,22,59].

Table 2. Observed Raman and Infrared Frequencies (cm^{-1}) for the $[KrF]^+$ Cation[a]

		ν(Kr—F)	ν(Kr ··· F)	δ(F—Kr ··· F)	Ref.
$[KrF]^+[Nb_2F_{11}]^-$	R[b]	613 (56) 606 (66) 597 (100)	372 (6)	194 (100) 189 (27) 181 (17.3) 167 (9.4)	23)
$[KrF]^+[Ta_2F_{11}]^-$	R[b]	609 (100) 600 (100) 594 (27)	337 (12)	Not obsd.	23)
$[KrF]^+[Sb_2F_{11}]^-$	R[c]	627 (100) 619 (20)	298 (6)? 270 (5) 260 (5)	150 (3)	23)
	I.r.[c]	616 m			23)
	R[d]	624 (100)	262 (6)	145 (4)	
$[KrF]^+[TaF_6]^-$	R[b]	603.5 (100) 599 (97)	343.5 (10) 339 (9) 325 (8.1)	192 (18) 179 (12)	23)
$[KrF]^+[PtF_6]^-$	R[d]	606 (50) 599 (60)	338 (3)	169 (3) 139 (8)	22)
$[KrF]^+[AuF_6]^-$	R[e]	597 (82)	346 (2)	163 (2)	21)
$\alpha[KrF]^+[AsF_6]^-$	R[d]	607 (100) 596 (100)	328 (12)	173 (8)	22)
$\beta[KrF]^+[AsF_6]^-$	R[d]	619 (72) 615 (100)	338 (16)	173 (10) 162 (11)	22)
$[KrF]^+[SbF_6]^-$	R[c]	621 (85) 618 (100)	348 (2.7) 344 (2.7)	174 (3.5) 166 (4.8) 149 (2.7)	23)
	I.r.[c]	607 s			23)
	R[d]	619 (74) 615 (100)	338 (4)	169 (5) 162 (7) 145 (3)	22)

R = Raman; I.r. = Infrared; m = medium; s = strong.
[a] Numbers in parentheses are relative intensities; [b] Spectra recorded at —196 °C; [c] Spectra recorded at room temperature; [d] Spectra recorded at —90 °C; [e] Spectra recorded at —80 °C.

So far no single-crystal X-ray structure determinations have been successfully carried out on any of the krypton difluoride-metal pentafluoride adducts; but X-ray work in the case of analogous xenon difluoride adducts has demonstrated that, in the solid state, the component molecules of the adducts are linked by fluorine bridges with a considerable degree of ionic character [30,34,37]. It is clear from Raman [18–20,22,23,59,60] infrared [19,21,23] and Mössbauer [24,25] studies that the krypton adducts are similar. Thus, observed Raman and infrared frequencies for the adducts are best assigned on the basis of ionic formulations, $[KrF]^+[M_2F_{11}]^-$, $[KrF]^+[MF_6]^-$ and $[Kr_2F_3]^+[MF_6]^-$, but the appearance of Kr ··· F and M ··· F bridge-stretching and F—Kr ··· F bending modes in the spectra are indicative of appreciable covalent character and contributions to the bonding from fluorine-bridged structures. For the $[KrF]^+[MF_6]^-$ adducts this interpretation is further substantiated by the more satisfactory assignments of the anion modes on the basis of C_{4v} rather than O_h symmetry [22,23]. The frequencies characteristic of $[KrF]^+$ and $[Kr_2F_3]^+$ in a variety of solid adducts together with their assignments are shown in Tables 2 and 3, respectively. The weighted means of the stretching frequencies associated with ν(Kr—F$^+$) in the series $[KrF]^+[M_2F_{11}]^-$ (M = Sb, Ta, Nb) decrease in the order Sb > Ta > Nb while those assigned to ν(Kr ··· F) increase [23]. Similar trends are observed in the $[KrF]^+[MF_6]^-$ and $[Kr_2F_3]^+[MF_6]^-$ series which are in accord with the accepted order of base strengths for the anions. Since ν(Kr—F) is highest for $[KrF]^+[Sb_2F_{11}]^-$ this can be regarded as most closely proximating to a salt [22,23]. The calculated $[Kr—F]^+$ stretching force constant is lower than that for $[Xe—F]^+$ and estimates of the bond lengths and force constants suggest a value of ~1.80 Å for Kr—F$^+$ which is longer than that predicted from an *ab initio* calculation [62]. The spectra of the $[Kr_2F_3]^+$ cation correlate well with those of the $[Xe_2F_3]^+$

Table 3. Observed Raman Frequencies (cm^{-1}) for the $[Kr_2F_3]^+$ Cation[a]

	ν(Kr_a—F_a)	ν(Kr_b—F_c)	ν(Kr_b ··· F_b)	ν(Kr_a ··· F_b)	δ(F—Kr ··· F)	Ref.
$[Kr_2F_3]^+[AsF_6]^-$ [b]	610 (43) 600 (80) 594 (100)	570 (4) 567 (31)	437 (5)	347 (sh) 336 (17)	239 (1) 183 (15) 174 (13) 158 (2)	22)
$[Kr_2F_3]^+[SbF_6]^-$ [b]	603 (100) 594 (89)	555 (34)	456 (4)	330 (18)	186 (16) 180 (sh) 176 (sh)	22)
	ν(Kr—F)		ν(Kr ··· F)			
$[Kr_2F_3]^+[SbF_6]^-$ [c]	607 (100) 602.5 (22.5) 595 (93) 587 (11) 565 (15) 559 (40)		334.5 (20) 325.5 (14)		δ(F—Kr ··· F) 190 (12.5) 186 (10) 181 (10.5) 175 (9)	23)
$[Kr_2F_3]^+[TaF_6]^-$ [c, d]	595 (60) 571 (30)					18, 23)

[a] Numbers in parentheses are relative intensities; [b] Spectra recorded at —90 °C; [c] Spectra run at —196 °C; [d] Spectrum run on frozen solution in BrF_5.

species [38,63]; but additional peaks, which have no equivalents in the xenon spectra, suggest that, unlike $[Xe_2F_3]^+$, the krypton cation is unsymmetrical (I) with one

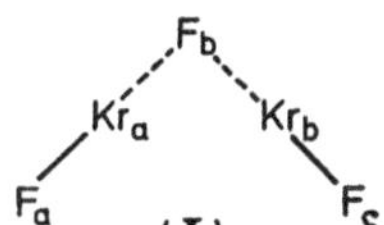

(I)

short and strong ($Kr_a—F_a$) bond, a weaker and longer ($Kr_b—F_c$) bond and two ($Kr_b \cdots F_b$) and ($Kr_a \cdots F_b$) bonds which are also of different lengths and strengths [22]. Another, but perhaps less satisfactory interpretation is that the cation may be regarded as a distorted KrF_2 molecule fluorine bridged to a $[KrF]^+$ ion, i.e. $[KrF_2 \cdot KrF]^+$ [19,22,23]. A definitive interpretation, however, must await a single-crystal X-ray diffraction study.

Two other interesting observations arise from the vibrational spectroscopy of the solid state. The first is that $[Kr_2F_3]^+$ has been shown to react further with KrF_2 to produce another species $[xKrF_2 \cdot Kr_2F_3]^+$ whose exact composition is still not certain [22]. The second is that during thermal decomposition studies of the tantalum and niobium adducts evidence for other adducts, which have been formulated as $[xKrF_2 \cdot KrF]^+$ (where x is probably equal to 1), has been observed [23]. In both cases it appears that additional KrF_2 units are weakly associated with the cations in the adducts.

Mössbauer spectra on α- and β-$[KrF]^+[AsF_6]^-$, $[KrF]^+[SbF_6]^-$ [13] and $[KrF]^+[AuF_6]^-$ [25] are in general agreement with the vibrational spectroscopic results, the parameters being accounted for in terms of the principal contributing valence bond structures (II_a–II_c). Although II_b and II_c produce no net additional changes in electric

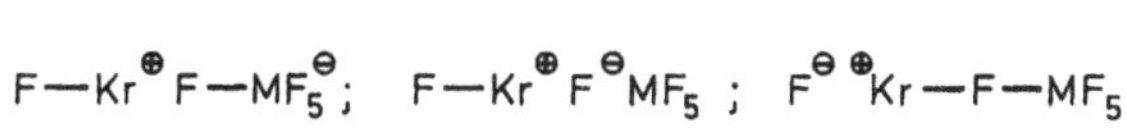

(IIa) (IIb) (IIc)

field gradient, II_a has the effect of elongating the $Kr \cdots F$ bridging bond and shortening the terminal one. It also places a single formal positive charge on the krypton. As the fluoride acceptor strength of the Lewis acid increases, II_a should be stabilized relative to II_b and II_c and quadrupole coupling should increase. Although in the case of $[KrF]^+[AsF_6]^-$ and $[KrF]^+[SbF_6]^-$ the trend is in the right direction, the experimental error is too high for it to be taken as significant [24].

The existence of the $[KrF]^+$ and $[Kr_2F_3]^+$ ions in solution has been determined by fluorine-19 n.m.r. and the parameters are listed in Table 4 [22]. The signal due to $[Kr—F]^+$, when KrF_2 dissolved in HF is added to excess of SbF_5, is observed to high field of that associated with KrF_2. This parallels the xenon case. In solution at low temperatures identical AX_2 spectra from BrF_5 solutions of $[Kr_2F_3]^+[AsF_6]^-$, $[Kr_2F_3]^+[SbF_6]^-$, as well as solutions containing KrF_2 and SbF_5 in a 2:1 mol ratio, can be unambiguously assigned to the symmetric V-shaped $[Kr_2F_3]^+$ cation rather than the unsymmetric species observed in the solid state [22]. In solution, therefore, it is clear that $[Kr_2F_3]^+$ and $[Xe_2F_3]^+$ are analogous. In other related fluoro- and oxyfluoro-cations of xenon characterized so far the fluorine-fluorine coupling con-

Table 4. ^{19}F N.m.r. Parameters for $[KrF]^+$ and $[Kr_2F_3]^+$ [22]

Solutes (m concn.)	Solvent	Temp. °C	J_{FF}, Hz: BrF_5	J_{FF}, Hz: $[Kr_2F_3]^+$	Chemical shift, ppm[a]: $[Kr_2F_3]^+$	Chemical shift, ppm[a]: $[KrF]^+$	Chemical shift, ppm[a]: KrF_2	Chemical shift, ppm[a]: Anion	Chemical shift, ppm[a]: Solvent
KrF_2 (0.22) SbF_5 (4.96)	HF	−40				22.6		92, 119, 139 $[Sb_2F_{11}]^-$ 124 $[SbF_6]^-$	192.4
$[Kr_2F_3]^+[AsF_6]^-$ (~0.5)	BrF_5	−65		347	A −19.0 X_2 −73.8			[b]	−150[b]
$[Kr_2F_3]^+[SbF_6]^-$ (~0.5)	BrF_5	−65		347	A −18.8 X_2 −73.4			115	−156[c]
KrF_2 (~1.0)[d] SbF_5 (~0.5)[d]	BrF_5	−66	75	351	A −19.0 X_2 −73.6			123	A −276.9 X_4 −137.0
KrF_2 (~1.5)[e] SbF_5 (~0.5)[e]	BrF_5	−65	75	351	A −19.2 X_2 −73.6		−65.5	120	A −276.4 X_4 −136.8

[a] With respect to external $CFCl_3$;
[b] The peak represents $[AsF_6]^-$ and BrF_5 fluorine environments undergoing rapid fluorine exchange;
[c] Axial and equatorial fluorine environments of BrF_5 collapsed into a single broad exchange-averaged peak;
[d] Solution after warming to room temperature for several seconds;
[e] KrF_2:SbF_5 calculated from the integrated KrF_2 and $[Kr_2F_3]^+$ peaks was 3:3:1.

stant is small (103–176 Hz)[43,64–66], and it has been suggested that the much larger values for $[Kr_2F_3]^+$ (349 Hz) and $[Xe_2F_3]^+$ (308 Hz)[43] may be associated with the fact that the F—Ng—F angle is ~180° in the $[Ng_2F_3]^+$ cations but only 90° in other cations[22]. Fluorine-19 n.m.r. spectra of BrF_5 solutions containing KrF_2:SbF_5 ratios of 3:3:1 gave no evidence for interactions between KrF_2 and $[Kr_2F_3]^+$ in solution[22].

The metal oxide tetrafluorides, $MoOF_4$ and WOF_4, are less powerful Lewis acids than most pentafluorides and, although the solid adducts, $KrF_2 \cdot MOF_4$ (M = Mo, W), have been prepared and $KrF_2 \cdot nMoOF_4$ (n = 1–3) and $KrF_2 \cdot WOF_4$ have been observed in solution at low temperature, and although at first sight these adducts appear to be similar to the pentafluoride complexes, Raman spectra of the solids and ^{19}F n.m.r. spectra of the solutions show that all the compounds are best formulated as essentially covalent structures containing fluorine bridges[49]. Because of the fact xenon difluoride adducts of WOF_4 exhibit fluorine bridge-oxygen bridge

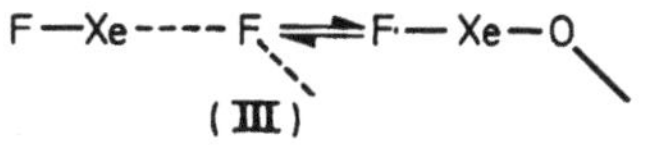

(III)

bond isomerization (III)[48] in solution, studies of similar low-temperature equilibria involving Kr—O—W and Kr ··· F ··· W-bridged species have been made and interpreted to suggest that stable krypton-oxygen bonds are unlikely. This is reinforced by the failure to prepare $Kr(OTeF_5)_2$[67] which would have been the most favourable case of a Kr—O bond, since the $OTeF_5$ group is almost as electronegative as F.

b Oxidative Fluorinating Abilities

While krypton difluoride will oxidize iodine to IF_7 and xenon to XeF_6[18] at room temperature, the oxidizing power of $[KrF]^+$ and $[Kr_2F_3]^+$ is even greater. Thus, IF_5 is readily oxidized to $[IF_6]^+$ by $[KrF]^+[Sb_2F_{11}]^-$[58], and $[KrF]^+$-containing adducts rapidly oxidize gaseous oxygen and xenon to $[O_2]^+$ and $[XeF_5]^+$[22]. Bromine pentafluoride solutions of $[KrF]^+[AsF_6]^-$[59], $[Kr_2F_3]^+[AsF_6]^-$ and $[Kr_2F_3]^+[SbF_6]^-$[20,59] at low temperatures initially give n.m.r. spectra of the cations. However, on warming, the BrF_5 is rapidly oxidized to $[BrF_6]^+$ (Eqs. 5 and 6):

$$BrF_5 + [KrF]^+ \rightarrow [BrF_6]^+ + Kr \quad (5)$$

$$BrF_5 + [Kr_2F_3]^+ \rightarrow [BrF_6]^+ + KrF_2 + Kr \quad (6)$$

Reaction of $[KrF]^+[SbF_6]^-$ or $[KrF]^+[Sb_2F_{11}]^-$ with excess of $XeOF_4$ has been shown to yield adducts containing $[XeOF_4 \cdot XeF_5]^+$ and $[O_2]^+$ salts probably according to the following Eqs. 7 and 8.

$$[KrF]^+ + 2\,XeOF_4 \rightarrow [XeOF_4 \cdot XeF_5]^+ + Kr + \frac{1}{2}O_2 \quad (7)$$

$$O_2 + [KrF]^+ \rightarrow [O_2]^+ + Kr + \frac{1}{2}F_2 \quad (8)$$

An earlier report[58] which suggested that $[XeOF_5]^+$ was the product was shown to be in error.

The $[KrF]^+[AuF_6]^-$ adduct, which can be prepared by the carefully controlled reaction of krypton difluoride with gold powder in anhydrous HF, has also been used in the synthesis of a number of other Au (V) species. Pyrolysis at 60–65 °C gives AuF_5, which itself will react with XeF_2 to produce $[Xe_2F_3]^+[AuF_6]^-$ and NOF to give $[NO]^+[AuF_6]^-$. The $[KrF]^+[AuF_6]^-$ adduct is also a potent oxidative fluorinating agent reacting rapidly with oxygen to give $[O_2]^+[AuF_6]^-$ and xenon to give $[XeF_5]^+$ [21].

4 $[XeF]^+$ and $[Xe_2F_3]^+$

a Adducts Incorporating Pentafluorides

Compounds of composition $XeF_2 \cdot 2\,MF_5$ (M = a pentavalent metal) have been known since 1963 [17,68,69], but it was not until the latter part of that decade that it was established that three classes of adducts of composition $XeF_2 \cdot 2\,MF_5$, $XeF_2 \cdot MF_5$ and $2\,XeF_2 \cdot MF_5$ existed [5,27–30,32,70,71] and the possibility of an even more extensive range was suggested by the reported isolation of $XeF_2 \cdot SbF_5$, $XeF_2 \cdot 1.5\,SbF_5$, $XeF_2 \cdot 2\,SbF_5$ and $XeF_2 \cdot 6\,SbF_5$ [27]. Both salt-like [32,70] and fluorine-bridged [17,69] formulations were suggested to explain the bonding in the compounds. By 1970, however, it was clear that, although the xenon difluoride-metal pentafluoride adducts have vibrational spectra which are best interpreted on the basis of formulations such as $[XeF]^+[M_2F_{11}]^-$, $[XeF]^+[MF_6]^-$ and $[Xe_2F_3]^+[MF_6]^-$, there is also considerable covalent character in the molecules and fluorine-bridged formulations contribute significantly to the bonding [29,30,32]. Early in the 1970's it was also established that the degree of ionic character in the adducts varies depending on the Lewis acid properties of the pentafluoride involved [33,38,43].

The range of adducts reported so far which conform to the $XeF_2:MF_5$, 1:2, 1:1 and 2:1 compositions are $XeF_2 \cdot 2\,NbF_5$ [29,35,36,38,41,42,72,73], $XeF_2 \cdot 2\,TaF_5$ [17,29,35,36,38,41,42,69,72,73], $XeF_2 \cdot 2\,RuF_5$ [29,32], $XeF_2 \cdot 2\,IrF_5$ [32], $XeF_2 \cdot 2\,SbF_5$ [8,17,30,31,33,38,40–42,44,45,71,73,74], $XeF_2 \cdot 2\,BiF_2$ [26], $XeF_2 \cdot VF_5$ [75,171], $XeF_2 \cdot NbF_5$ [29,35,36,38,41,42,72,73], $XeF_2 \cdot TaF_5$ [29,35,36,38,40–42,72,73], $XeF_2 \cdot RuF_5$ [29,32,34], $XeF_2 \cdot OsF_5$ [29,32], $XeF_2 \cdot IrF_5$ [32,76], $XeF_2 \cdot PtF_5$ [32], $XeF_2 \cdot AsF_5$ [28,32,33,39,43,45,76,77], $XeF_2 \cdot SbF_5$ [27,33,36,41–45,71], $XeF_2 \cdot BiF_5$ [26,40], $2\,XeF_2 \cdot TaF_5$ [36,38,41], $2\,XeF_2 \cdot RuF_5$, $2\,XeF_2 \cdot OsF_5$ [32], $2\,XeF_2 \cdot IrF_5$ [32,76], $2\,XeF_2 \cdot PtF_5$ [32], $2\,XeF_2 \cdot AuF_5$ [21], $2\,XeF_2 \cdot AsF_5$ [32,33,37,40,45,70,76,77], $2\,XeF_2 \cdot SbF_5$ [33,36,38,41,43] and $2\,XeF_2 \cdot BiF_5$ [26].

The above complexes are normally prepared by fusing together appropriate amounts of the component fluorides under an atmosphere of dry, inert gas. Sometimes it is convenient to add one component in excess and to remove the unreacted reagent under vacuum until a constant weight is obtained. Preparations have also been carried out in solutions of non-reducing solvents such as HF or BrF_5 at room temperature or with gentle warming if solution is not easily effected at ambient temperature. It appears that preparation in solution is the only route for fluoroarsenate derivatives [28,32,33,37,39,40,43,45,70,76,77]. Very few other preparative methods have been reported. However, $[O_2]^+[SbF_6]^-$ [30], $[Cl_2F]^+[AsF_6]^-$ [77] and $[BrF_6]^+[AsF_6]^-$ [59], have all been shown to oxidize xenon to give $[XeF]^+$ or $[Xe_2F_3]^+$ adducts, and xenon and fluorine react spontaneously in the dark with liquid antimony pentafluoride to give $[XeF]^+[Sb_2F_{11}]^-$ [8].

Single crystal structure determinations have been carried out on representatives of each of the above 1:2, 1:1 and 2:1 classes above namely, those of $XeF_2 \cdot 2\,SbF_5$ ($[XeF]^+[Sb_2F_{11}]^-$) [30,31], $XeF_2 \cdot RuF_5$ ($[XeF]^+[RuF_6]^-$) [34], $XeF_2 \cdot AsF_5$ ($[XeF]^+[AsF_6]^-$ [39]), and $2\,XeF_2 \cdot AsF_5$ ($[Xe_2F_3]^+[AsF_6]^-$ [37]). The crystal structure data are given below and the molecular geometries are shown in Fig. 1.

$[XeF]^+[Sb_2F_{11}]^-$ [30,42)f] Very pale yellow crystals, monoclinic, $a = 8.07\,(1)$, $b = 9.55\,(1)$, $c = 7.33\,(1)$ Å, $\beta = 105.8\,(1)°$, $V = 543$ Å^3, $Z = 2$, $d_c = 3.69$ g cm^{-3}. Space group $P2_1$.
Xe—F_t = 1.82 (3) Å, Xe—F_b = 2.34 (3) Å, F—Xe—F = 176.1 (1.4)°, Xe—F—Sb = 149 (2)°.

$[XeF]^+[RuF_6]$ [34)f] Very pale green crystals, monoclinic, $a = 7.991\,(6)$, $b = 11.086\,(6)$, $c = 7.250\,(6)$ Å, $\beta = 90.68\,(5)°$, $V = 642.2$ Å^3, $Z = 4$, $d_c = 3.78$ g cm^{-3}. Space group $P2_1/n$.
Xe—F_t = 1.872 (17) Å, Xe—F_b = 2.182 (15) Å, F—Xe—F = 177.08 (1.2)°, Xe—F—Ru = 137.19 (46)°.

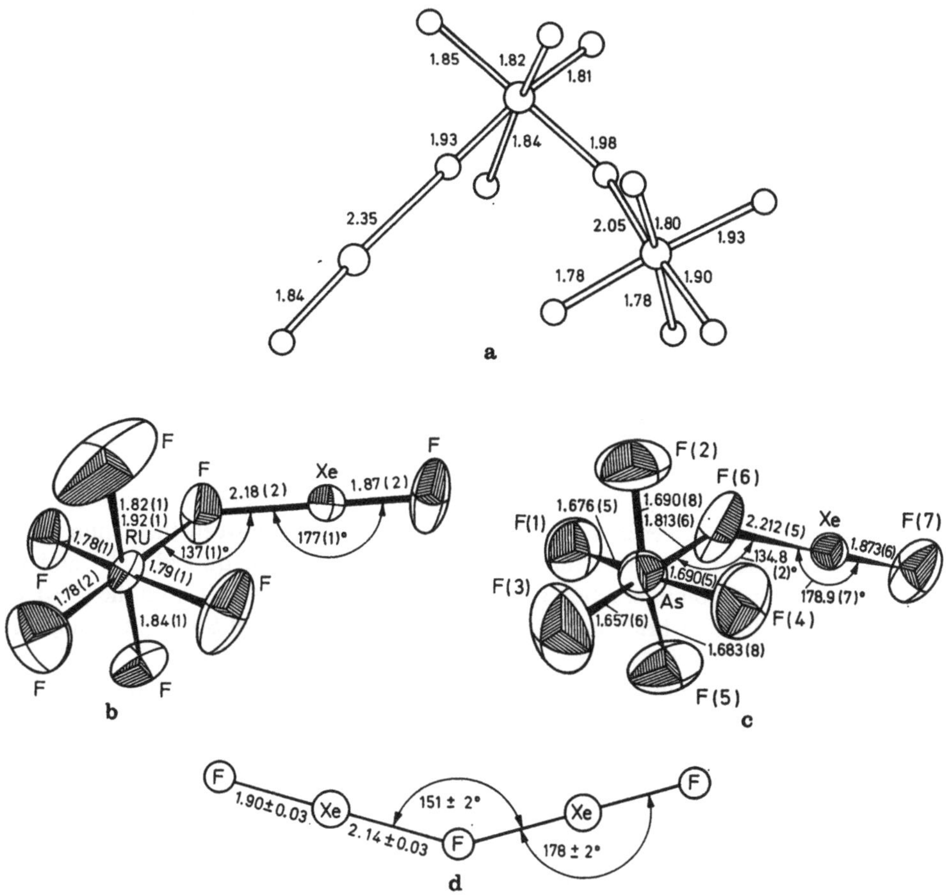

Fig. 1a—d. Molecular Geometries of Xenon Difluoride-Metal Pentafluoride Adducts. **a** The formula unit of $XeFSb_2F_{11}$; **b** The formula unit of $XeFRuF_6$; **c** The formula unit of $XeFAsF_6$; **d** The $[Xe_2F_3]^+$ cation; (distances in Ångstoms and angles in degrees throughout).

$[XeF]^+[AsF_6]^-$ [39)f] Very pale yellow-green crystals, monoclinic, $a = 6.308$ (3), $b = 6.275$ (3), $c = 16.023$ (5) Å, $\beta = 99.97$ (5)°, $V = 624.66$ Å^3, $Z = 4$, $d_c = 3.61$ g cm^{-3}. Space group $P2_1/n$.
$Xe—F_t = 1.873$ (6) Å, $Xe—F_b = 2.212$ (5) Å, F—Xe—F = 178.9 (7)°, Xe—F—As = 134.8 (2)°.
$[Xe_2F_3]^+[AsF_6]$ [37,70)f] Very pale yellow crystals, monoclinic, $a = 15.443$ (10), $b = 8.678$ (5), $c = 20.888$ (15) Å, $\beta = 90.13$ (6)°, $V = 2799.3$ Å^3, $Z = 12$, $d_c = 3.62$ g cm^{-3}. Space group $I2/a$.
Within the two crystallographically non-equivalent $[Xe_2F_3]^+$ ions, which are not significantly different from each other, the average values are as follows: $Xe—F_t = 1.90$ (3) Å, $Xe—F_b = 2.14$ (3) Å, F—Xe—F = 178 (2)°, Xe—F—Xe = 150.2 (8)°. The bridging F atom of each cation has two short contacts at ~3.0 Å to terminal fluorines on each of two other cations. The nearest contacts of the $[AsF_6]^-$ anion are at 3.4 Å from the bridging fluorine of the $[Xe_2F_3]^+$ ion.

The structures are similar in that all contain similar $Xe—F_t$ bond lengths, and Fe—Xe—F and Xe—F—M or Xe—F—Xe bond angles. The $Xe—F_b$ bond lengths vary from 2.34–2.14 Å over the four structures, the longest being that associated with $[XeF]^+[Sb_2F_{11}]^-$ [30,42)]. Although this structure, therefore, is derived from an ionic formulation, study of the molecular geometry (Fig. 1) and the crystal structure suggests that such a formulation grossly over-simplifies the type of interactions in the crystal. Although the $Xe—F_t$ bond length of 1.82 Å is less than those in XeF_2 (2.0 Å) the proximity of the other fluorine at 2.34 Å is evidence of considerable interaction between the xenon and one of the Sb_2F_{11} fluorine atoms [30,31)]. The similarity of the geometries and bond lengths in the $FXeFRuF_5$ [34)] and $FXeFAsF_5$ [39)] units indicates that the bonding is essentially the same for both. Although it has been strongly argued that in each case $FXeFMF_5$ should be represented as the salt $[XeF]^+[MF_6]^-$, with the short distance between the xenon of cation and one of the fluorines of the anion arising as a consequence of the cation having its positive charge centered largely at the xenon atom [34,39)], it is quite clear that, as in $FXeFSb_2F_{10}$, important contributions to the bonding arise from a rather strong fluorine bridge with, presumably, considerable covalent character. The crystal structure of $2\,XeF_2 \cdot AsF_5$ indicates that the $[Xe_2F_3]^+[AsF_6]^-$ formulation, with a V-shaped $[Xe_2F_3]^+$ cation is appropriate [37,70)]. However, no Xe ··· F contact distance between the xenon atoms in $[Xe_2F_3]^+$ and the nearest fluorine in $[AsF_6]^-$ has been quoted [37,70)] and the extent of cation-anion interaction is therefore difficult to assess. Within the $[Xe_2F_3]^+$ ion itself the fluorine bridging angle is close to that in $FXeFSb_2F_{10}$ although the average $Xe—F_b$ bond is shorter.

If the $[Xe_2F_3]^+$ cation is considered in terms of an assemblage of two $[XeF]^+$ and one F^- ions, each $[XeF]^+$ species receives electron density from the associated F^- and the assembly becomes a multicentre bonded system with a lower overall electron affinity than that of $[XeF]^+$ [78)]. The measured fluoride ion donor ability of xenon difluoride, ($\Delta H^0 = 9.45$ eV) [14)], is greater than that of XeF_4 but less than that of XeF_6 (see Section 2a, p. 65). The increase in Xe—F bond energy in cation formation [195.9 kJ mol^{-1} in $[XeF]^+$ c.f. 135.4 kJ mol^{-1} in XeF_2 [14)]] contributes to

[f] t = terminal; b = bridging.

this fluoride ion donor ability. As evidenced by the X-ray work, modest fluoride ion donor ability is thus a feature of the chemistry of XeF_2, with the cations, $[XeF]^+$ and $[Xe_2F_3]^+$, forming one or more fluorine bridges to the anion. This overall picture of bonding is confirmed by vibrational, Mössbauer and nuclear quadrupole resonance spectroscopic work and by thermochemical studies.

Raman and infrared spectra are most readily assigned on the basis of ionic formulations $[XeF]^+[M_2F_{11}]^-$, $[XeF]^+[MF_6]^-$ and $[Xe_2F_3]^+[MF_6]^-$, but modes associated with Xe ··· F and M ··· F bridge-stretching and F—Xe ··· F bending confirm the occurrence of fluorine bridges in the structures. Additional corroborative evidence comes from observation of the spectra associated with the anions in the $[XeF]^+[MF_6]^-$ adducts. Although early assignments in some cases were made on the basis of the anion having O_h symmetry [33] more than 6 anion modes are usually observed and a more satisfactory assignment is usually achieved if the anion is regarded as distorted from O_h to C_{4v} symmetry. This, for example, avoids a requirement to break the mutual exclusion rule with respect to v_3 in $[XeF]^+[SbF_6]^-$ [26,38], and v_1 in $[XeF]^+[TaF_6]^-$ and $[XeF]^+[NbF_6]^-$ [38] and factor group splitting of v_1 in $[XeF]^+[SbF_6]^-$ [26,38].

The frequencies characteristic of $[XeF]^+$ and $[Xe_2F_3]^+$ in a variety of adducts, together with their assignments, are shown in Tables 5 and 6 respectively. Since v(Xe—F) is totally symmetric its multiplicity of bands in many spectra is attributed to factor group splitting [33,38]. Variation in the average values of the v($[Xe—F]^+$) stretching frequency for series of the adducts (see Table 5 and Fig. 2) has been attributed to progressive strengthening of this bond with increasing Lewis acidity of the associated pentafluoride [38].

The crystal structure determination on $[Xe_2F_3]^+[AsF_6]^-$ [37,70] has demonstrated unequivocally that the cation is V-shaped, with C_{2v} symmetry. On the basis of there

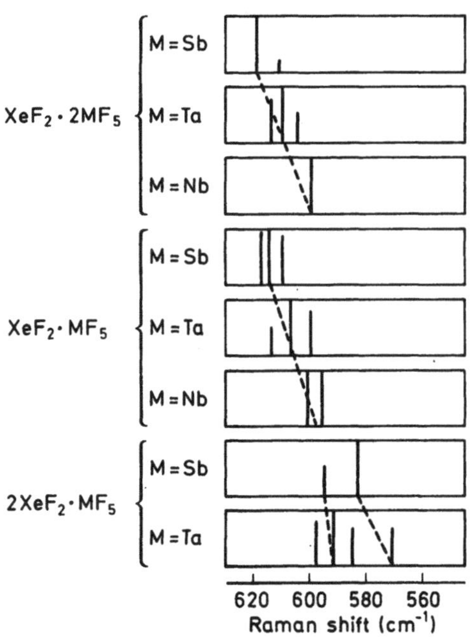

Fig. 2. Stretching Frequencies Associated with v(Xe—F)

Table 5. Observed Raman and Infrared Frequencies (cm^{-1}) for the $[XeF]^+$ Cation[a]

		ν(Xe—F)	ν(Xe ··· F)	δ(F—Xe ··· F)	Ref.
$[XeF]^+[Nb_2F_{11}]^-$	R	600 (100)	297 (5)	152 (3)	[38]
	I.r.	600 m			[38]
$[XeF]^+[Ta_2F_{11}]^-$	R	605 (75)	294 (6)	115 (5)	[38]
		610 (100)		147 (10)	
		614 (80)			
	I.r.	608 br, s			[38]
$[XeF]^+[Ru_2F_{11}]^-$	R	598 (53)			[32]
		604 (100)			
$[XeF]^+[Ir_2F_{11}]^-$	R	601 (68)			[32]
		612 (100)			
$[XeF]^+[Sb_2F_{11}]^-$	R	611 sh(18)	271 (8)	131 (15)	[38]
		619 (100)			
	I.r.	611 w			[38]
		626 w			
$[XeF]^+[NbF_6]^-$	R	596 (100)		123 (9)	[38]
		691 (95)		151 (12)	
				193 (6)	
$[XeF]^+[TaF_6]^-$	R	600 (83)	318 (45)	122 (9)	[38]
		607 (100)		179 (4)	
		614 (53)			
	I.r.	603 s			[38]
$[XeF]^+[RuF_6]^-$	R	599 (86)			[32]
		604 (51)			
$[XeF]^+[IrF_6]^-$	R	602 (60)[c]			[32]
		608 (44)			
$[XeF]^+[PtF_6]^-$	R	602 (35)			[32]
		609 (23)			
$[XeF]^+[AsF_6]^-$	R	609 (100)	339 (21)	99 (1)	[33]
				146 sh(8)	
				155 (15)	
				180 (2)	
	I.r.	565 w			[28]
$[XeF]^+[SbF_6]^-$	R	610 sh(98)		140 (22)	[38]
		615 (100)			
		617 (100)			
	I.r.	608 m			[38]
$[XeF]^+[BiF_6]^-$	R[b]	602 (48)		123 (9)	[26]
		608 (11)		151 (12)	
				193 (6)	

R = Raman; I.r. = Infrared; w = weak; m = medium; s = strong; sh = shoulder.
[a] Numbers in parentheses are relative intensities; [b] Spectra recorded at −196 °C; [c] May be due to $[XeF]^+[IrF_6]^-$ impurity.

being nine Raman-active vibrational modes (Eq. 9), rigorous assignments of the shifts

$$\Gamma = 4a_1 + 3b_1 + a_2 + b_2 \qquad (9)$$

observed in the low-temperature Raman spectra of three of the $[Xe_2F_3]^+$ adducts (see Table 6) have been made [26]. These data are of significantly better quality

Table 6. Observed Raman and Infrared Frequencies (cm^{-1}) for the $[Xe_2F_3]^+$ Cation[a]

		ν(Xe—F)	ν(Xe ··· F)	δ(F—Xe···F)	Ref.
$[Xe_2F_3]^+[TaF_6]^-$	R	571 (70)	428 br (2)	155 (13)	[38]
		585 (73)		164 (25)	
		592 (100)		182 (19)	
		598 (85)			
	I.r.	586 w			[38]
		592 w			
		600 w			
$[Xe_2F_3]^+[RuF_6]^-$	R	579 (140)		171 (30)	[32]
		593 (200)		159 (30)	
$[Xe_2F_3]^+[OsF_6]^-$	R	582 (190)		161 (90)	[32]
		593 (130)		169 sh	
$[Xe_2F_3]^+[IrF_6]^-$	R	578 (123)		154 (19)	[32]
		592 (73)		160 (34)	

		$\nu_5(b_1)$ asym. Xe—F str.	$\nu_1(a_1)$ sym. Xe—F str.	$\nu_6(b_1)$ asym. Xe···F str.	$\nu_2(a_1)$ sym. i.p. F—Xe···Fδ	$\nu_7(b_1)$ asym. i.p. F—Xe···Fδ	$\nu_8(a_2)$ sym. o.p. F—Xe···Fδ	$\nu_9(b_2)$ asym. o.p. F—Xe···Fδ	$\nu_3(a_1)$ sym. Xe···F str.	$\nu_4(a_1)$ sym. F···Xe···Fδ	Ref.
$[Xe_2F_3]^+[AsF_6]^-$	R[b]	598 (95)	588 (100)	417 (<1) 401 (<1)	255 br (5)			163 (37)			[26,35]
$[Xe_2F_3]^+[SbF_6]^-$	R[b]	591 (66)	582 (100)	420 (<1)	255 (<1)	179 (6)	171 (15)	161 (21)			[26,33]
	I.r.	595 m	570 m								[38]
$[Xe_2F_3]^+[BiF_6]^-$	R[b]	589 (60)	580 (73)	388 (<1)	248 (4)		171 (8)	160 (9)	118 (2)	86 (2)	[26]

R = Raman; I.r. = Infrared; w = weak; m = medium; sh = shoulder.
[a] Numbers in parentheses are relative intensities; [b] Spectra recorded at —196 °C.
i.p. = ‚in-plane'; o.p. = 'out-of-plane'.

than the earlier room-temperature data and probably provide the best interpretation of the spectra so far.

The observed quadrupole coupling constants obtained in ^{93}Nb n.q.r. studies on $[XeF]^+[NbF_6]^-$ (34.6 MHz at 77 K) is much greater than would be expected from a structure involving $[NbF_6]^-$ ions since, in the latter case, only a small field gradient is expected [35]. It has been concluded, therefore, that one of the niobium-fluorine bonds is altered through participation in a fluorine bridge. N.q.r. data used to predict coupling constants for $[XeF]^+[Nb_2F_{11}]^-$ give excellent agreement with observed values for the *cis* structure [35] and strongly suggests that the structure approximates to that of $[XeF]^+[Sb_2F_{11}]^-$ [30,31].

Measurements of quadrupole couplings and isomer shifts from ^{129}Xe Mössbauer spectra of $[XeF]^+[TaF_6]^-$, $[XeF]^+[BiF_6]^-$, $[XeF]^+[Sb_2F_{11}]^-$, $[Xe_2F_3]^+[AsF_6]^-$ and $[Xe_2F_3]^+[BiF_6]^-$ have not given such definitive results. The lack of accuracy of the isomer shifts made it impossible to derive detailed conclusions. The range of shifts ($s = -0.1 \rightarrow 0.5$ mm s^{-1}) overlaps those for other classes of Xe^{II} and Xe^{IV} species and no evidence for systematic dependence of the isomer shift on chemical parameters within any of the groups has been found. However, a slight increase of the quadrupole interaction strength with acceptor strength of the associated pentafluoride is evident so, again, interaction between cation and anion via fluorine bridging is implied [40].

Evidence of variation in ionic character in series of the adducts in the solid state has been detected in both vibrational and Mössbauer spectroscopic results. In the Raman spectra of 1:2, 1:1 and 2:1 series of XeF_2—MF_5 adducts (M = Sb, Ta or Nb) variation in the average values of $\nu([Xe—F]^+)$ show that the adduct with the greatest ionic character is the antimony-containing compound and the fluorine bridging becomes increasingly significant when antimony is replaced first by tantalum and then niobium [38]. This is in accord with accepted Lewis acid strengths of the pentafluorides involved. The temperatures required for thermal decomposition also reflect the same trend [38] and, with the exception of the reaction of XeF_2 with excess of SbF_5 (which yields $[XeF]^+[Sb_2F_{11}]^-$), the temperature necessary for complete reaction of the components in the formation of the adducts decreases in the order Sb > Ta > Nb for each series [38]. These three trends parallel values of the enthalpies of formation of the adducts from their components derived from measurements of heats of alkaline hydrolysis (Table 7) [36]. Thus the values for $[XeF]^+[Sb_2F_{11}]^-$ (-59 kJ mol^{-1}) and $[XeF]^+[SbF_6]$ (-6 J kJ mol^{-1}) are modestly negative and these two have the greatest empirical thermal stability and are closest to having salt like structures. The equivalent niobium compounds, on the other hand, are least ionic and decompose at temperatures little above room temperature [29,38]. The trends in enthalpy of formation for the three series of adducts suggest that the value for $[Xe_2F_3]^+[NbF_6]^-$ should be significantly positive (i.e. $> +50$ kJ mol^{-1}) and this is in keeping with the failure to prepare this adduct under ordinary conditions [36,38].

D.t.a. investigations showed that chemical equilibria in the XeF_2—MF_5 systems are established rather slowly [73], and later weight-loss versus time-of-pumping studies on the products of reaction of large excesses of xenon difluoride with the pentafluorides of antimony, tantalum and niobium showed that, in addition to the 1:2, 1:1 and 2:1 compounds, species with compositions $[2\,XeF_2 \cdot Xe_2F_3]^+[SbF_6]^-$,

Table 7. 2:1, 1:1 and 1:2 Adducts of XeF_2 and MF_5 (M = Nb, Ta, Sb) Comparison of Enthalpies of Hydrolysis, ΔH_h;[a] Enthalpies of Reaction of Component Fluorides, ΔH,[b] and Specific Conductivities at their Melting Points [31, 36, 42]

Adduct	ΔH_h (kJ mol^{-1})	ΔH (kJ mol^{-1})	Specific conductivity $\times 10^4$ (ohm^{-1} cm^{-1})[e]
$[Xe_2F_3]^+[NbF_6]^-$ [c]		> +50	
$[Xe_2F_3]^+[TaF_6]^-$	−1,109[d]	> +50	
$[Xe_2F_3]^+[SbF_6]^-$	−1,172[d]	> + 8	
$[XeF]^+[NbF_6]^-$	− 773	+38	42.5
$[XeF]^+[TaF_6]^-$	− 711	−17	49.4
$[XeF]^+[SbF_6]^-$	− 769	−67	146.5
$[XeF]^+[Nb_2F_{11}]^-$	−1,194[d]	+50	10.6
$[XeF]^+[Ta_2F_{11}]^-$	−1,124[d]	− 4	25.2
$[XeF]^+[Sb_2F_{11}]^-$	−1,174	−59	35.9

[a] = m aqueous alkali at 298.2 K; [b] = at 298 K; [c] = not preparable under ordinary conditions; [d] = limiting value; [e] = at the melting point.

$[XeF_2 \cdot XeF]^+[TaF_6]^-$ and $[XeF_2 \cdot XeF]^+[NbF_6]^-$, which contain loosely bound xenon difluoride, are also formed [38].

In the molten state and in solution the picture is much the same as in the solid state. The most striking feature of the Raman spectra of melts of $[XeF]^+[MF_6]^-$ and $[XeF]^+[M_2F_{11}]^-$ is the persistence of peaks associated with Xe ··· F and M ··· F bridging bonds [41]. However, in general, the bonds associated with $\nu([XeF]^+)$ are shifted slightly towards lower and $\nu([M{-}F]^-)$ towards higher frequencies than the average values for the solids as expected due to the increase in thermal energy. The spectra due to $[XeF]^+[Sb_2F_{11}]^-$ exhibit no evidence of XeF_2 or SbF_5 in the melt, but in the case of the tantalum and niobium analogues free pentafluoride is observed. This suggests that an equilibrium (Eq. 10):

$$[XeF]^+[M_2F_{11}]^- \rightleftharpoons XeF_2 + 2\,MF_5 \qquad (M = Ta\ or\ Nb) \tag{10}$$

albeit, heavily displaced towards the left, occurs. Melt spectra of $[XeF]^+[SbF_6]^-$ indicate that neither dissociation nor disproportionation occur in the melt. In the analogous tantalum and niobium cases, however, slight dissociation occurs and an equilibrium (Eq. 11):

$$3\,[XeF]^+[MF_6]^- \rightleftharpoons XeF_2 \cdot [XeF]^+[MF_6]^- + [XeF]^+[M_2F_{11}]^- \tag{11}$$

is implied, With $[Xe_2F_3]^+[MF_6]^-$ (M = Sb, Ta) significant changes in the spectra occur on melting and this has been interpreted in terms of an equilibrium (Eq. 12):

$$\begin{array}{c} [Xe_2F_3]^+\,[SbF_6]^- \rightleftharpoons [XeF_2 \cdot XeF]^+\,[SbF_6]^- \\ \qquad\qquad\qquad\qquad \upharpoonleft\!\downharpoonright \\ \qquad\qquad\qquad XeF_2 + [XeF]^+\,[SbF_6]^- \end{array} \tag{12}$$

Another view of the nature of the adducts in the molten state has been gained from measurement of the specific conductivities at the melting point (Table 7). The values are in the range $10–150 \times 10^{-4}$ ohm^{-1} cm^{-1}, and are of the same order as those of alkyl ammonium picrates, which are known to be incompletely dissociated. It is also noteworthy that the least conducting have specific conductivities only a little higher than those of molten pentafluorides [42]. For $[XeF]^+[SbF_6[^-$, which has the highest specific conductivity, the electrical conductivity measurement together with a viscosity determination has yielded a value of the order of 11% for the degree of dissociation near the melting point [42].

In solution valuable data have come from ^{19}F [43] and ^{129}Xe [44,45] n.m.r. spectra. The ^{19}F chemical shifts and coupling constants are summarized in Table 8. A single fluorine resonance is found for the $[XeF]^+$ cation but the chemical shift and $^{129}Xe—^{19}F$ coupling constant differ depending on the solvent. The observation that there is a continuous change in the chemical shift with changing relative amounts of HSO_3F and SbF_5 solvent has been interpreted in terms of the strengths of the bonding of the $[XeF]^+$ to the accompanying anions. The $[XeF]^+$ cation is assumed to be solvated by forming a single additional bond with the most basic (strongest donor) molecule present. Increase in chemical shift and $^{129}Xe—^{19}F$ coupling constant has been correlated with $[XeF]^+$ becoming more weakly bonded to the less basic accompanying anions. Thus, in solutions of $[XeF]^+[SbF_6]^-$ or $[XeF]^+[Sb_2F_{11}]^-$ in fluorosulphuric acid, if SbF_5 is added, as the SbF_5 concentration is increased the HSO_3F solvent is replaced by $[Sb_2F_{11}]^-$ and $[SbF_5SO_3F]^-$ and, in turn, by the more weakly basic $[Sb_nF_{5n+1}]^-$ anions [43]. The ^{129}Xe chemical shift of $[XeF]^+$ has also been shown to be very sensitive to solvent composition and, as with the ^{19}F chemical shift, there is continuous change as $HSO_3F:SbF_5$ ratio is changed. The explanation is also the same. More interestingly, ^{129}Xe Fourier Transform n.m.r. spectroscopic work has shown that ^{129}Xe chemical shift data provides a very sensitive probe for assessing the degree of ionic character in the Xe—F bond. Plots of ^{129}Xe chemical shift versus ^{19}F chemical shift of the terminal fluorine on xenon exhibit a linear

Table 8. ^{19}F Nmr Parameters for the $[XeF]^+$ and $[Xe_2F_3]^+$ Cations [43]

Solute (molal conc^n)		Solvent^a	Temp. °C	Chem. Shift^b ppm		$J_{^{129}Xe-F}$ Hz	$J_{FF'}$ Hz
$[XeF]^+[SbF_6]^-$	(0.70)	HSO_3F	−93	$[XeF]^+$	242.5	6620	
				$[SbF_6]^-$	123.2		
$[XeF]^+[AsF_6]^-$	(0.46)	HSO_3F	−96	$[XeF]^+$	243.8	6615	
				$[AsF_6]^-$	64.4		
XeF_2^{c}	(1.57)	SbF_5	26	$[XeF]^+$	291.9	7210	
				$[Sb_nF_{5n+1}]^-—SbF_5$	115.1		
XeF_2^{d}	(0.34)	SbF_5	26	$[XeF]^+$	289.8	7210	
				$[Sb_nF_{5n+1}]—SbF_5$	111.1		
$[Xe_2F_3]^+[AsF_6]^-$		BrF_5	−62	$[Xe_2F_3]^+$	184.7 A	4865	308
					252.0 X_2	6740	
				$[AsF_6]^-$	61.8		

^{a} HSO_3F, −40.8 ppm; BrF_5, −272.8 (F_a) and −134.3 (F_e) ppm, J_{FF} = 75.2 Hz; ^{b} With respect to external $CFCl_3$; ^{c} Bright yellow solution; ^{d} Green solution.

relationship over the whole range of Xe^{II} species and the covalency of the bond increases with decreasing ^{129}Xe chemical shift and increasing ^{19}F chemical shift. The $[XeF]^+$ cation in SbF_5 solution is clearly the closest to being a free cation [45]. The $[Xe_2F_3]^+$ cation, which can be regarded in terms of the contributing valence-bond structures F—Xe—F^+Xe—F, F—Xe^+F—Xe—F, and F—Xe^+F^- $^+$Xe—F, together with those for XeF_2, would be expected to have a terminal Xe—F bond order and a corresponding ^{129}Xe chemical shift intermediate between those of XeF_2 and $[XeF]^+$, and this exactly what is found [45].

In addition to the adducts described above, a number of other pentafluoride adducts of similar composition, $XeF_2 \cdot 2\,BrF_5$ [73], $XeF_2 \cdot 2\,IF_5$ [79], $XeF_2 \cdot PF_5$ [80], $XeF_2 \cdot IF_5$ [73,81] and $2\,XeF_2 \cdot PF_5$ [80], have been prepared by direct combination of the XeF_2 and MF_5 components under suitable conditions. A single-crystal X-ray structure determination on $XeF_2 \cdot IF_5$ [81] has shown that this complex is a molecular adduct in which the molecular dimensions of the constituents are not significantly different from those of the pure components. Structure determinations on $XeF_2 \cdot XeF_4$ [82] and $XeF_2 \cdot XeOF_4$ [83] have shown that these, too, are similar. It is assumed that this whole range of compounds, together with $XeF_2 \cdot 1.5\,IF_5$, $XeF_2 \cdot 9\,MF_5$ (M = Br, I) [73], together with $XeF_2 \cdot [XeF_5]^+[AsF_6]^-$ and $XeF_2 \cdot 2\,([XeF_5]^+[AsF_6]^-)$ [83], all incorporate molecular xenon difluoride and should not be confused with the adducts described earlier.

b Adducts Incorporating Tetrafluorides and Hexafluorides

Infrared spectra of 1:1 adducts of xenon difluoride with molybdenum and tungsten hexafluorides and 2:1 adducts with tin, zirconium and hafnium tetrafluorides have been assigned on the basis of their containing $[XeF]^+$ cations $[XeF]^+[MF_6]^{2-}$ (M = Ti, Zr, Hf) [84,85]. This is supported in the case of $2\,XeF_2 \cdot SnF_4$ by Mössbauer work [86]. Data on $nXeF_2 \cdot TiF_4$ (n = 1.5, 1 and 0.5) are less well defined [87].

c Adducts Incorporating Oxide Tetrafluorides

Thermally stable adducts involving fluoride acceptors in the weak to intermediate range of fluoride-acceptor strengths have been prepared only quite recently. The solid adducts $XeF_2 \cdot WOF_4$ and $XeF_2 \cdot 2\,WOF_4$ were prepared by interaction of stoichiometric amounts of XeF_2 and WOF_4 [46]. On the basis of low-temperature ^{19}F [46] and ^{129}Xe [45,48,49] studies in BrF_5 and SO_2ClF solvents and Raman spectroscopic studies of the solids [46,49] it has been shown that the tungsten compounds [45,46,48,49] and the analogous molybdenum oxide tetrafluoride adducts [45,48,49] are fluorine-bridged to xenon and can be formulated as shown (IV and V). A single-crystal X-ray study [47]

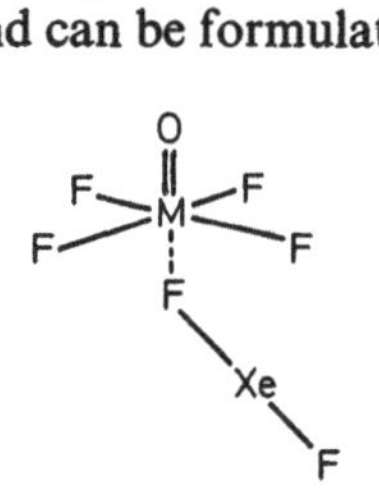

(IV)

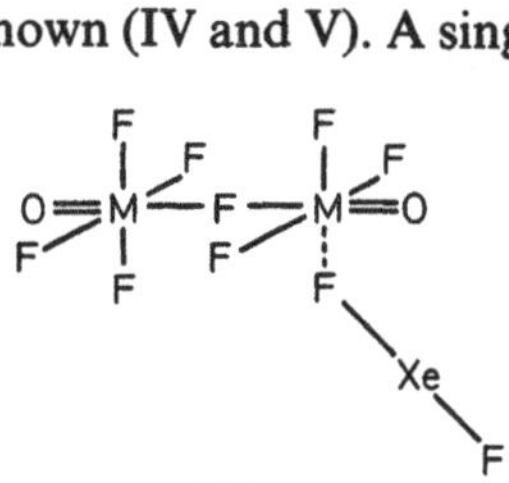

(V)

has shown that $XeF_2 \cdot WOF_4$ has approximately C_s symmetry, with the terminal Xe—F bond length (1.89 Å) shorter and the Xe ··· F bridge bond length (2.04 Å) slightly longer than the Xe—F bond length in XeF_2 (2.00 Å), thus confirming the solid-state Raman spectroscopic results.

Study of the ^{19}F and ^{129}Xe n.m.r. spectra of the $XeF_2 \cdot MOF_4$ and $XeF_2 \cdot 2\,MOF_4$ (M = Mo or W) species in SO_2ClF solvent has resulted in observation of equilibria involving higher chain length species $XeF_2 . nMOF_4$ (n = 1–4; M = Mo or W)[45,48]. In all of these solution species fluorine exchange involving dissociation of the Xe ··· F bridge bond is slow, thus indicating that the Xe ··· F ··· M bridge is non-labile in solution on the n.m.r. time scale[46,48]. Isomerization between oxygen and fluorine-bridged XeF groups has been observed in the tungsten adducts, $XeF_2 \cdot nWOF_4$ n = 2 and 3, a phenonemon which had not been previously observed in noble-gas chemistry. The relative degree of covalent character in the terminal Xe—F bonds and the relative fluoride ion acceptor strengths of the $(MoOF_4)_n$ and $(WOF_4)_n$ (n = 1–4) polymer chains have been assessed from the ^{19}F and ^{129}Xe n.m.r. complexation shifts. In all cases it is clear that ionic character in these adducts is small[48].

d Adducts Incorporating Fluorosulphate and Related Groups

One or both of the fluorines in XeF_2 may be substituted by other highly electronegative groups. The preparation of the monosubstituted derivatives generally involve interaction of the fluoride with the appropriate anhydrous acid (Eq. 13), the elimination of the highly exothermic HF providing

$$XeF_2 + HL \rightarrow FXe—L + HF \qquad (13)$$

the driving force for the reactions. The first compounds of this type to be prepared were xenon (II) fluoride fluorosulphate, $FXeSO_3F$, and xenon (II) fluoride perchlorate, $FXeClO_4$[99], but now a number of monosubstituted derivatives are known. The substituted ligands include $—OC(O)CF_3$[89], $—OPOF_2$[90], $—OSO_2F$[43,45,88,91], $—OSeF_5$[92,93], $—OTeF_5$[93–95] and $—OClO_3$[88]. The pentafluorotellurate is the most stable[94], but the perchlorate and the trifluoroacetate are kinetically unstable and detonate readily[88]. The X-ray single crystal structure of the fluorosulphate (Fig. 3) has established that the compound exists in the solid state as discrete $FXeOSO_2F$ molecules[88,91]. As in the case of $XeF_2 \cdot WOF_4$ the terminal Xe—F bond length (1.94 Å) is shorter than in XeF_2 (2.00 Å). The bridging Xe ··· F bond length is

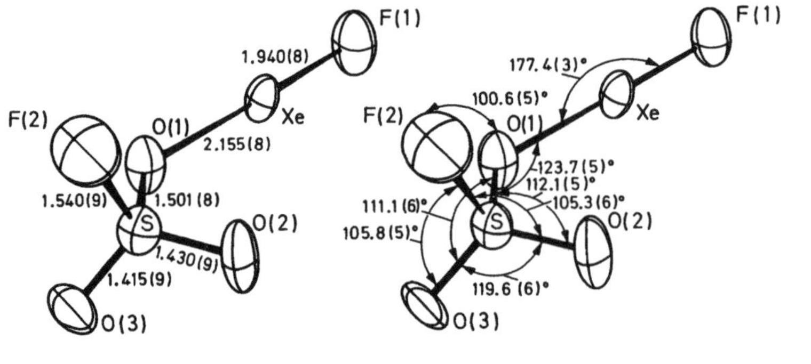

Fig. 3. Molecular Geometry of $FXeOSO_2F$

somewhat longer (2.16 Å) than that in XeF_2 and, in fact, is intermediate between that in XeF_2 and that in $[XeF]^+[Sb_2F_{11}]^-$ (2.35 Å). The inference is that $FXeOSO_2F$ is essentially covalent, with little contribution to the bonding from structures such as $[F—Xe]^+[OSO_2F]^-$, and it is assumed that the related compounds will be similar. In solution the structure is also clearly the same. A near-linear correlation exists between the ^{19}F chemical shift and the $^{129}Xe—^{19}F$ coupling constant for a series of F–Xe ··· containing compounds and $FXeSO_3F$ exhibits the lowest chemical shift and coupling constant, while "free" $[XeF]^+$ cation in SbF_5 solution has the highest values [43,45].

e $FXeN(SO_2F)_2$

The first report of the preparation of $FXeN(SO_2F)_2$ [96] caused excitement, because it was the first indication that xenon might bond to elements other than fluorine or oxygen. An X-ray single-crystal study has shown that the solid consists of $FXeN(SO_2F)_2$ molecules of molecular point symmetry C_2 in which the xenon atom of the Xe—F group is bonded to the nitrogen (Fig. 4) [97]. Use of the bond order-bond length relationship of Pauling for calculation of the terminal Xe—F bond order has resulted in the conclusion that valence bond structure (VI) has only a 59:41 dominance over (VII) (Eq. 14).

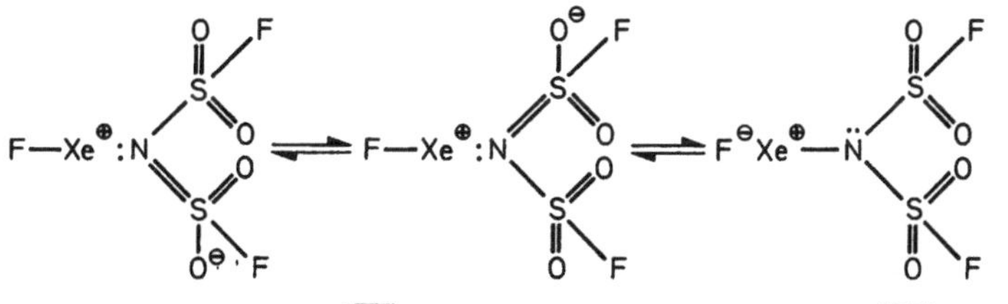

(VI) (VII)

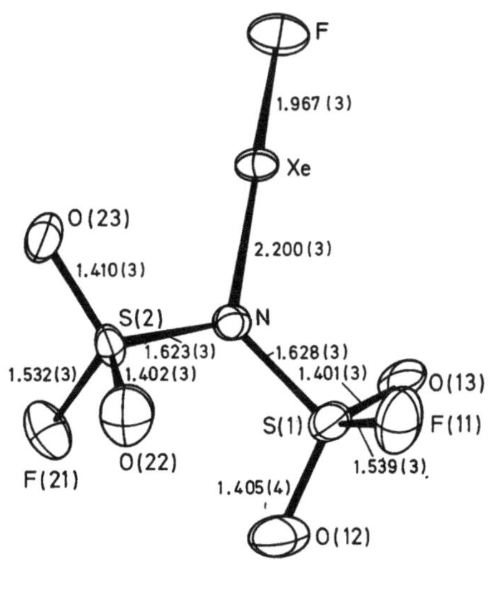

Fig. 4. Molecular Geometry of $FXeN(SO_2F)_2$

The Xe—F stretching mode in the Raman of the solid at 506 cm^{-1} is the lowest value of ν(Xe—F) observed in any F—Xe—Y species so far [97], being less than those observed in $FXeOSO_2F$ (528 cm^{-1}) [33] and $FXeOTeF_5$ (520 cm^{-1}) [95]. In view of the very small shift from the value of the XeF stretching frequency in XeF_2 (497 cm^{-1}) the amount of $[Xe—F]^+$ character must be very small. A similar conclusion can be drawn from ^{19}F and ^{129}Xe data. The similarity of $FXeN(SO_2F)_2$ and XeF_2 chemical shifts relative to those in other xenon (II) compounds is a clear indication that the bonding in the F—Xe—N moiety is similar to that in XeF_2 itself [45,97].

f $XePtF_6$ and Related Compounds

It was the preparation of $[O_2]^+[PtF_6]^-$ [98] which led to the discovery of the first noble gas compound to be reported. To this xenon species was ascribed the formula $[Xe]^+[PtF_6]^-$ [99], and is formed spontaneously at room temperature by the interaction of xenon gas with PtF_6 [32,68,99]. Work with other hexafluorides [32,100] has shown that thermodynamically unstable ones (PtF_6, RuF_6, RhF_6 and PuF_6) react with xenon at room temperature to give compounds of the type $Xe(MF_6)_x$ (x = 1–2), while stable hexafluorides (e.g. UF_6, OsF_6, IrF_6 and NpF_6) do not react under ordinary conditions. Infrared data leave little doubt that the platinum and rhodium compounds contain Pt(V) and Rh(V). Their reaction with alkali fluoride in IF_5 also certainly generates $A^IM^VF_6$ salts. If the compounds contain $[Xe]^+$ then they should be paramagnetic. However, so far, no reliable magnetic data have been obtained [32,68]. Regrettably, because '$XePtF_6$' is amorphous it has been impossible to compare powder photographs with those of $[XeF]^+[PtF_6]^-$ [32]. Consequently, it is still not certain whether the solid compounds designated $[Xe]^+[PtF_6]^-$ and $Xe(PtF_6)_2$ and their relatives contain $[Xe]^+$, $[Xe_2]^{2+}$ or $[Xe_2F]^+$ or whether "$XePtF_6$" is in fact $XePtF_7$ and "$XePtF_6$" and $Xe(PtF_6)_2$ are the compounds $[XeF]^+[PtF_6]^-$ and $[XeF]^+[Pt_2F_{11}]^-$.

5 $[Rn]^{2+}$, $[RnF]^+$ and $[Rn_2F_3]^+$

The early characterization of a low-volatility radon-fluorine compound by combination of radon with fluorine at 400 °C [68,101] was followed by work which showed that ^{222}Rn is oxidized by chlorine and bromine fluorides, IF_7 [102–105] or $[NiF_6]^{2-}$ [103–105] in HF to give stable solutions of radon fluoride. On the basis of electromigration studies it has been suggested that the radon in these solutions is present as $[RnF]^+$ or $[Rn]^{2+}$ [104,105], but $[Rn_2F_3]^+$ also seems a likely possibility. The application of relativistic quantum mechanics to radon fluoride structures also shows that ionic formulations are favoured [106].

No complex salts containing $[RnF]^+$, $[Rn_2F_3]^+$ or $[Rn]^{2+}$ have yet been isolated and fully characterized. However, it has been shown that radon can be collected by oxidation of gaseous radon with liquid BrF_3 and the solid complexes $ClF_2 \cdot SbF_6$, $BrF_2 \cdot SbF_6$, $BrF_4 \cdot Sb_2F_{11}$, $IF_4(SbF_6)_3$ and $BrF_2 \cdot BiF_6$. The radon is probably being converted to $[RnF]^+$ by reactions along the following lines [107–108] (Eq. 15–19):

$$Rn_{(g)} + [ClF_2]^+[SbF_6]^-_{(s)} \rightleftharpoons [RnF]^+[SbF_6]^-_{(s)} + ClF_{(g)} \quad (15)$$

$$Rn_{(g)} + [BrF_2]^+[SbF_6]^-_{(s)} \rightarrow [RnF]^+[SbF_6]^-_{(s)} + BrF_{(g)} \quad (16)$$

$$Rn_{(g)} + [BrF_4]^+[Sb_2F_{11}]^-_{(s)} \rightarrow [RnF]^+[SbF_6]^-_{(s)} + [BrF_2]^+[SbF_6]^-_{(s)} \quad (17)$$

$$Rn_{(g)} + [IF_4]^{3+}[SbF_6]^-_{3(s)} \rightarrow [RnF]^+[SbF_6]^-_{(s)} + [IF_4]^+[Sb_2F_{11}]^-_{(s)} \quad (18)$$

$$Rn_{(g)} + [BrF_2]^+[BiF_6]^-_{(s)} \rightarrow [RnF]^+[BiF_6]^-_{(s)} + BrF_{(g)} \quad (19)$$

More recently the case for the existence of $[RnF]^+$ has been strengthened by a study of the reaction of radon with solid $[O_2]^+[SbF_6]^-$ at 25 °C [74]. A non-volatile radon species is produced and, since the analogous reaction with xenon has been shown to yield gaseous oxygen and $[XeF]^+[Sb_2F_{11}]^-$, it is assumed that the radon reaction is as follows (Eq. 20):

$$Rn_{(g)} + 2\,([O_2]^+[SbF_6]^-{}_{(s)} \rightarrow [RnF]^+[Sb_2F_{11}]^-_{(s)} + 2\,O_{2(g)} \quad (20)$$

III $[XeF_3]^+$

1 Introduction

Of the three binary xenon fluorides, XeF_4 appears to have the most limited scope of complex formation. The only adducts which have been isolated at room temperature are $XeF_4 \cdot SbF_5$ [58,109,110], $XeF_4 \cdot 2\,SbF_5$ [58,65], $XeF_4 \cdot BiF_5$ and $XeF_4 \cdot 2\,BiF_5$ [26] This paucity of complex formation has been used to grade the relative fluoride ion donor abilities in the order $XeF_6 > XeF_2 \gg XeF_4$ and was exploited as a chemical method for the purification of XeF_4 [76]. There are, however, differences of opinion as to the relative fluoride ion donor abilities of XeF_2 and XeF_4 [4].

2 Syntheses

$XeF_4 \cdot SbF_5$ and $XeF_4 \cdot 2\,SbF_5$ are prepared by heating (80 °C) mixtures of the components, one of which is added in excess [58,109,110]. After removal of excess of SbF_5, $XeF_4 \cdot 2\,SbF_5$ is obtained as a pale yellow-green solid (m.p. 81–83 °C). The 1:1 adduct can be obtained as a pale yellow solid (m.p. 109–113 °C) by fusion of an excess of XeF_4 with the latter. However, the compound is dimorphic and a second form can be prepared from SbF_5 and an excess of XeF_4 in anhydrous HF [111]. A pale yellow-green solid is crystallized after removing excess of solvent and XeF_4. These have been designated as the high-temperature (80 °C) α-form and the room temperature β-form. They can be distinguished by their Raman spectra which are similar, but each line of the $[XeF_3]^+$ ion in the β-form is split into a doublet (see Table 9).

Because of the low volatility of BiF_5, the 1:1 and 1:2 XeF_4:BiF_5 adducts were prepared in anhydrous HF by mixing appropriate stoichiometric amounts of the components.

Evidence has been adduced for the existence of an adduct, $XeF_4 \cdot AsF_5$, stable only at low temperatures in the presence of an excess of AsF_5 [111].

Table 9. Observed Raman Frequencies (cm^{-1})[a] and Assignments for $[XeF_3]^+$ in $[XeF_3]^+[MF_6]^-$

$[XeF_3]^+$ $[BiF_6]^-$ [b]	α-$[XeF_3]^+$ $[SbF_6]^-$ [c]	β-$[XeF_3]^+[SbF_6]^-$ [c]	$[XeF_3]^+[AsF_6]^-$ [c]	Assignment
645 (100)	643 (100)	663 (100), 643 (56)	643 (85)	$\nu_1(a_1)$ $\nu(Xe{-}F_e)$
609 (10)	609 (9)	604 (21), 612 (25)	607 (38), 608 (58)	$\nu_4(b_1)$ $\nu_{asym}(Xe{-}F_a)$
557 (83)	573 (88)	564 (94), 576 (94)	571 (100)	$\nu_2(a_1)$ $\nu_{sym}(Xe{-}F_a)$
374 (2)	305 sh	318 (2), 335 (2)	316 (18)	$\nu_5(b_1)$ $\delta_{asym}(F_a{-}Xe{-}F_a)$
253 (3)				$\nu_6(b_2)$ $\delta(XeF_3)$, out of plane
198 (7)	205 (2)	199 (2), 212 (3)		$\nu_3(a_1)$ $\delta_{sym}(F_a{-}Xe{-}F_a)$

[a] Numbers in parentheses are relative intensities; [b] Ref. [26]; [c] Ref. [111].

3 Structures

Characterization of the structures of these complexes by a variety of physical means has shown that they can be described in terms of the T-shaped cation, $[XeF_3]^+$, coordinated to anions via fluorine bridges the lengths and polar character of which are largely determined by the Lewis acidity of the anion.

The most unambiguous descriptions of $[XeF_3]^+$ in the solid state have been obtained by crystal structure determinations of $[XeF_3]^+[SbF_6]^-$ [112], $[XeF_3]^+[Sb_2F_{11}]^-$ [58,113] and $[XeF_3]^+[BiF_6]^-$ [114]. The crystal data are given below and the molecular geometries are shown in Fig. 5.

$[XeF_3]^+[SbF_6]^-$ [112] Monoclinic, $a = 5.394$ (1), $b = 15.559$ (2), $c = 8.782$ (1) Å, $\beta = 103.10$ (1)°, $V = 717.84$ Å^3, $Z = 4$, $d_c = 3.92$ g cm^{-3}. Space group $P2_1/n$.

$[XeF_3]^+[Sb_2F_{11}]^-$ [113] Triclinic, $a = 8.237$ (5), $b = 9.984$ (20), $c = 8.004$ (5) Å, $\alpha = 72.54$ (5), $\beta = 112.59$ (7), $\gamma = 117.05$ (21)°, $V = 534.9$ Å^3, $Z = 2$, $d_c = 3.98$ g cm^{-3}. Space group $P\bar{1}$.

$[XeF_3]^+[BiF_6]^-$ [114] Triclinic, $a = 5.698$ (3), $b = 7.811$ (3), $c = 8.854$ (4) Å, $\alpha = 99.45$ (5), $\beta = 110.09$ (5), $\gamma = 92.84$ (3)°, $V = 362.7$ Å^3, $Z = 2$, $d_c = 4.69$ g cm^{-3}. Space group $P\bar{1}$.

The three structures have, of course, a number of common features. The $[XeF_3]^+$ cation has the T-shaped structure expected for an AX_3E_2 molecule; i.e. a trigonal bipyramidal geometry, the two non-bonding electron pairs occupying equatorial positions to minimize interactions with bonding electron pairs [115]. Structural features are rather similar to those of the isoelectronic molecules BrF_3 and ClF_3.

In $[XeF_3]^+[SbF_6]^-$ the closest bridging distance to a fluorine of the anion is ~2.49 Å, indicating weak covalent interaction. This fourth fluorine atom in the xenon coordination sphere is coplanar with the other three fluorines bonded to the xenon. In order to avoid electron pairs in the xenon valence shell the closest approach of the nearest bridged fluorine atom is towards the middle of the triangular planes of the trigonal bipyramid. An additional, relatively short, intermolecular contact of 2.71 Å with an adjacent $[XeF_3]^+[SbF_6]^-$ unit links two of these into what are essentially dimeric units [112].

The bond lengths and angles in the $[XeF_3]^+$ units in both the $[XeF_3]^+[SbF_6]^-$ [112] and $[XeF_3]^+[Sb_2F_{11}]^-$ [113] structures are rather similar as are the short contacts of

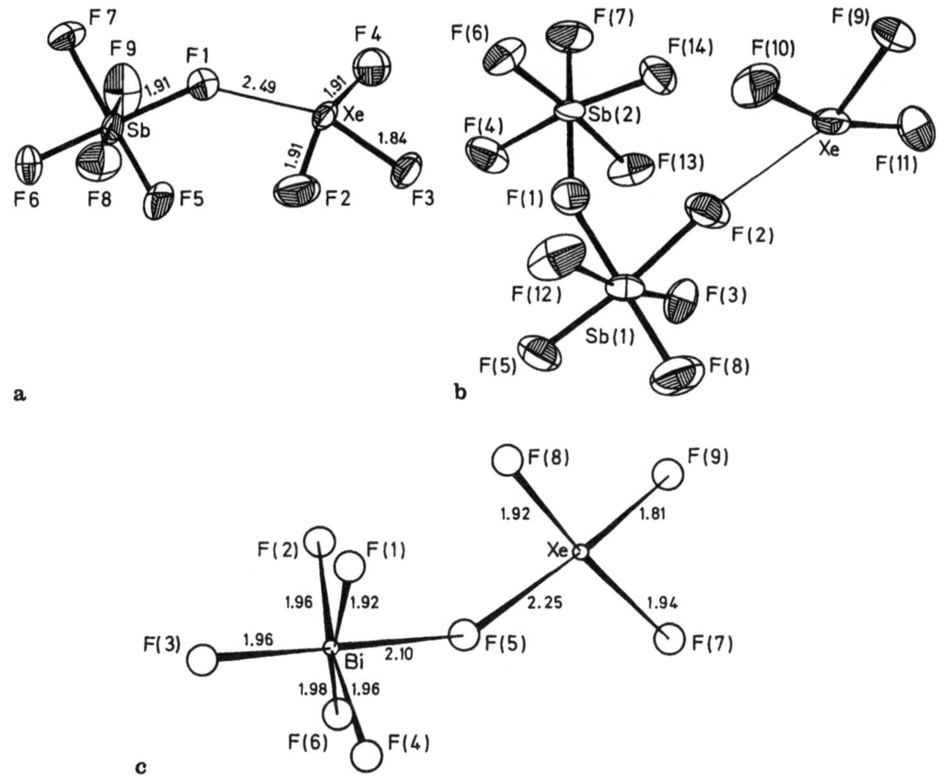

Fig. 5a—c. Molecular Geometries of Xenon Tetrafluoride-Metal Pentafluoride Adducts. **a** A perspective view of the $XeF_3^+SbF_6^-$ structural unit; **b** The $[XeF_3^+][Sb_2F_{11}^-]$ structural unit; **c** The $[XeF_3]$-$[BiF_6]$ structural unit

~2.50 Å between xenon and a fluorine on the $[SbF_6]^-$ (*viz.* $[Sb_2F_{11}]^-$) anion. The major difference between the two is in the bridge-bond angle which is 140.8° for the 1:1 compound and 171.6° for the 1:2 compound. The most likely explanation for this is packing considerations. It is noteworthy that the departure from the strictly ionic salt formulation causes the bridging Sb—F distance in $[XeF_3]^+[SbF_6]^-$ to be longer than the other Sb—F bond distances; i.e. a departure from octahedral symmetry for $[SbF_6]^-$. As outlined below, this is reflected in the increased complexity of the Raman spectra.

The structure of $XeF_4 \cdot BiF_5$ can also be described in terms of an approximate $[XeF_3]^+$ cation linked to a $[BiF_6]^-$ anion. Structural parameters for $[XeF_3]^+$ are similar to those for the SbF_5 adducts. However, in $[XeF_3]^+[BiF_6]^-$ [114] there is a rather short contact (2.25 Å) between the xenon and the fluorine linking it via a fluorine bridge to bismuth. At the same time, while in the SbF_5 adducts the incoming electron pair from the fluorine is directed towards the centre of the triangular face of the $[XeF_3]^+$ trigonal pyramid in order to avoid electrons in the xenon valence shell, in the BiB_5 adducts, this lone pair is directed towards the middle of an edge of this trigonal bipyramid due to the strong fluorine bridge-xenon interaction. This con-

Table 10. Comparison of Quadrupole Splittings (ΔE_Q) and Isomer Shifts (S) for some Xe(II) and Xe(IV) Compounds, derived from ^{129}Xe Mössbauer Spectra (Ref. [40])

	ΔE_Q (mm s^{-1})	S (mm s^{-1})
XeF_4	41.04 (7)	0.40 (4)
$[XeF_3]^+[Sb_2F_{11}]^-$	39.6 (1)	0.20 (6)
$[XeF_3]^+[BiF_6]^-$	41.3 (1)	0.30 (5)
XeF_2	39.7 (4)	0.2 (2)
$[XeF]^+[Sb_2F_{11}]^-$	41.5 (2)	0.2 (1)
$[XeF]^+[BiF_6]^-$	41.4 (1)	0.20 (5)

sequence of the lower Lewis acidity of BiF_5 thus leads to a structure which is intermediate between an ionic $[XeF_3]^+[BiF_6]^-$ and a molecular adduct $XeF_4 \cdot BiF_5$.

Mössbauer spectra [40] have been obtained for quadrivalent xenon compounds including the complexes $[XeF_3]^+[Sb_2F_{11}]^-$ and $[XeF_3]^+[BiF_6]^-$. In these the quadrupole splitting is barely affected by their degree of covalency. However, both the isomer shifts and the quadrupole splittings, although sharply distinguishable from Xe(VI) compounds, are rather similar to those of Xe(II) compounds (Table 10). This technique, therefore, has contributed relatively little to the characterization of these complexes.

Evidence for the existence of T-shaped $[XeF_3]^+$ in solution has been obtained by nuclear magnetic resonance spectroscopy on both ^{19}F [65, 110] and ^{129}Xe [44, 45] nuclei. The ^{19}F n.m.r. spectra show the characteristic AB_2 spectrum expected for this structure and confirmation of the T-shaped structure in SbF_5 solution was confirmed by ^{129}Xe n.m.r. work. Data are given in Table 11.

Table 11. ^{19}F and ^{129}Xe Nuclear Magnetic Resonance Parameters for $[XeF_3]^+$

	Chemical Shift (ppm)		Coupling Constants (Hz)	
	$^{19}F^a$	$^{129}Xe^b$	J_{FF}	$J_{^{129}Xe-^{19}F}$
A	−23.0[c] (T)	595[d]	174[c]	2440[c], 2384[d] (D)
B_2	−38.7[c] (D)			2620[c], 2609[d] (T)

[a] With respect to external $CFCl_3$; [b] With respect to external $XeOF_4$; [c] From ^{19}F data, Ref. [110]; [d] From ^{129}Xe data, Refs. [44,45].

4 Vibrational Spectroscopy

The normal modes for a T-shaped $[XeF_3]^+$ cation of C_{2v} symmetry span the following irreducible representations: $3a_1 + 2b_1 + b_2$, all of which are infrared and Raman active. Only Raman measurements have been made [26,109,111], and these mostly on solid compounds. Assignments of stretching frequencies were complicated by the overlap of cation and anion frequencies in the 450–750 cm^{-1} region of the spectrum. Moreover, the $[SbF_6]^-$ anion does not exhibit a simple spectrum, the relatively strong fluorine bridging causing severe distortion from O_h symmetry. In fact, for $[XeF_3]^+[SbF_6]^-$ and $[XeF_3]^+[AsF_6]^-$, the anion spectra have been interpreted in terms of C_{4v} symmetry resulting from this distortion. Assignments are aided by comparison with the isoelectronic halogen trifluorides, those of the symmetric Xe—F_{axial} and Xe—$F_{equatorial}$ stretching frequencies being particularly straightforward, aided by polarization measurement in SbF_5 solution [109]. Fuller assignments were made later [26] and these are given in Table 9. The cationic lines associated with β-$[XeF_3]^+[SbF_6]^-$ are each split into a doublet. Since these are not degenerate, the splitting cannot be due to site symmetry and they have been attributed, therefore, to factor-group splitting [111].

IV $[XeF_5]^+$ and $[Xe_2F_{11}]^+$

1 Introduction

The cationic derivatives of xenon hexafluoride are among the most thoroughly investigated species in noble gas chemistry. The possible existence of $[XeF_5]^+$ as a stable species was inferred early on because of the high conductivities of XeF_6/HF solutions as compared to those of XeF_2 or XeF_4 in HF [118]. The following equilibrium was suggested (Eq. 21).

$$XeF_6 + HF \rightarrow [XeF_5]^+ + [HF_2]^- \quad (21)$$

Although the $[XeF_5]^+$ cation exists in HF solution, it was later shown that solution of xenon hexafluoride in HF involves a more complex equilibrium (Eq. 22 and 23) [117] than was initially thought.

$$([XeF_5]^+F^-)_4 + nHF \rightleftharpoons 2\,[Xe_2F_{11}]^+ + [(HF)_nF]^- \quad (22)$$

$$[Xe_2F_{11}]^+ + nHF \rightleftharpoons 2\,[XeF_5]^+ + [(HF)_nF]^- \quad (23)$$

The ionic nature of these species was deduced from the very beginning on the basis of the identification of the anionic counter ion, $[BF_4]^-$, in the complex $XeF_6 \cdot BF_3$ [116], but the first characterization of the $[XeF_5]^+$ cation itself was only attempted later [170]. The full characterization of $[Xe_2F_{11}]^+$ was even more recent [120]. The two hexavalent xenon cations are related in much the same way as $[XeF]^+$ and $[Xe_2F_3]^+$ in xenon(II) complexes; i.e. the simpler cations are joined by a fluorine bridge.

Numerous adducts of XeF_6 with fluoride ion acceptors are now known. These include stoichiometries such as 1:2, 1:1, 2:1 and others. Complexes reported are $XeF_6 \cdot 2\,AlF_3$ [121], $XeF_6 \cdot 2\,SbF_5$ [40,64,122], $[XeF_5]^+[SO_3F]^-$ [64,123], $XeF_6 \cdot FeF_3$ [124], $XeF_6 \cdot BF_3$ [64,116,117,125], $XeF_6 \cdot GaF_3$, $XeF_6 \cdot InF_3$ [121], $XeF_6 \cdot CrF_4$ [126], $XeF_6 \cdot VF_5$ [127], $XeF_6 \cdot NbF_5$ [128], $XeF_6 \cdot TaF_5$ [129], $XeF_6 \cdot RuF_5$ [34,117], $XeF_6 \cdot IrF_5$ [76],

$XeF_6 \cdot PtF_5$ [76,117,170,130], $XeF_6 \cdot AuF_5$ [117], $XeF_6 \cdot UF_5$ [131–133], $XeF_6 \cdot AsF_5$ [37,64,76,116,117,125,134], $XeF_6 \cdot SbF_5$ [64,122], $2\,XeF_6 \cdot PdF_4$ [117,135], $2\,XeF_6 \cdot GeF_4$ [136], $2\,XeF_6 \cdot SnF_4$ [137], $2\,XeF_6 \cdot VF_5$ [127], $2\,XeF_6 \cdot TaF_5$ [129], $2\,XeF_6 \cdot RuF_5$, $2\,XeF_6 \cdot IrF_5$ [76], $2\,XeF_6 \cdot PtF_5$ [119], $2\,XeF_6 \cdot AuF_5$ [120], $2\,XeF_6 \cdot PF_5$ [117,119,134], $2\,XeF_6 \cdot AsF_5$ [76,117,134] and $2\,XeF_6 \cdot SbF_5$ [122] as well as more complex stoichiometries such as $4\,XeF_6 \cdot MnF_3$, $4\,XeF_6 \cdot MnF_4$ [129], $4\,XeF_6 \cdot PdF_4$ [120], $4\,XeF_6 \cdot GeF_4$ [136], $4\,XeF_6 \cdot SnF_4$ [134] and $mXeF_6 \cdot MF_4 (m \leqq 6;\ M = Zr, Hf)$ [138].

2 $[XeF_5]^+$

a The Preparation and Nature of the Adducts

The complexes are normally prepared by melting the appropriate proportions of the constituents and pumping off any excess of reactant until constant weight is attained. In some cases the complexes can be prepared by mixing the reactants in solutions of non-reducing solvents such as HF or BrF_5. Several have been prepared by fluorinating lower-valent fluorides in the presence of an excess of XeF_6 (e.g. $[XeF_5]^+[PtF_6]^-$ [130]. The adduct $[XeF_5]^+[SO_3F]^-$ has been obtained from the reaction of XeF_6 with HSO_3F [64,123].

The stoichiometries of these complexes are most likely determined by the tendency of XeF_6 towards formation of stabilized cations containing $[XeF_5]^+$ moieties and of the anions towards formation of octahedral hexafluoro-anions, $[MF_6]^-$ (viz. $[MF_6]^{2-}$). The list of XeF_6 complexes is large, but we shall refer here only to those where more than passing reference is made to their salt-like nature.

Although there is disagreement as to the relative fluoride ion donor abilities of XeF_2 *vis-a-vis* XeF_4, there is no doubt that XeF_6 is by far the strongest fluoride ion donor [4,76]. Electron spectroscopy, ^{19}F nuclear magnetic resonance and Mössbauer spectroscopy all indicate considerable migration of electrons Xe → F in the Xe—F bond in the binary xenon fluorides [2]. According to the three-centre four-electron bond model for the binary xenon fluorides [139,140] the net charge per fluorine atom in all cases is approximately —0.5. Thus, the approximate bond polarities in these compounds may be written $Xe^{+}(F^{-0.5})_2$, $Xe^{+2}(F^{-0.5})_4$ and $Xe^{+3}(F^{-0.5})_6$; i.e., increased positive charge on the central xenon atom with increased oxidation number. The net charge on the xenon atom in reality is not this large by far, as can be seen from observed trends in the ^{129}Xe chemical shifts. Nevertheless, it would be expected that fluoride acceptor strengths would increase in the order $[XeF]^+ < [XeF_3]^+ < [XeF_5]^+$. Viewed differently, the fluoride ion donor abilities of the fluorides would decrease in the order $XeF_2 > XeF_4 > XeF_6$. The actual relative fluoride ion donor abilities in terms of the enthalpies of ionization as derived from photoionization studies [14] are given in Table 12 [83]. The enthalpy of ionization of XeF_6 is thus about 0.6 eV less than expected from extrapolation of the XeF_2 and XeF_4 values [83]. This has been ascribed to the energetically favored pseudo-octahedral structure of $[XeF_5]^+$ [4,76]. Indeed, pseudo-potential SCF-MO calculations for $[XeF_5]^+$ and XeF_6 have shown that their structures are governed by the stereochemically active lone pairs [119]. For XeF_6, crowding of six ligands and a lone pair would favour loss of an F^- anion. For $[XeF_5]^+$ there is close agreement between the calculated (80.8°) and observed (79.2°) F_{eq}—Xe—F_{ax} (F_{eq} = equatorial F; F_{ax} = axial F) bond angles, but calculated

Table 12. Enthalpies of Ionization of the Xenon Fluorides as Derived From Photoionization Studies [14]

Ionization Process	ΔH° (eV)
$XeF_{2(g)} \rightarrow [XeF]^+_{(g)}$	9.45
$XeF_{4(g)} \rightarrow [XeF_3]^+_{(g)} + F^-_{(g)}$	9.66
$XeF_{6(g)} \rightarrow [XeF_5]^+_{(g)} + F^-_{(g)}$	9.24

differences, $r(Xe—F_{eq}) - r(Xe—F_{ax}) = 0.005$ compare less favourably with the observed difference of 0.03 ± 0.02 Å. This may be due, however, to the fact that calculations were done for the bare ion while observed values were obtained from possibly distorted solid-state environments. An additional consequence of the high positive charge on the xenon(VI) species is the appreciable polarizing capability of the cation. This is reflected in the formation of fluorine bridges to the anions causing some distortion from octahedral symmetry.

The unique role of the $[XeF_5]^+$ cation in XeF_6 chemistry is reflected in the structure of the parent molecule, solid XeF_6. Crystalline XeF_6 is polymorphic, existing in four phases. Only the cubic phase has been described in detail [141,142] (Fig. 6). The structure is based on the association of $[XeF_5]^+$ and F^- ions into tetrameric and hexameric units. The shape of the $[XeF_5]^+$ ion conforms with that expected from VSEPR theory, and is similar in size and shape in both the hexamers and tetramers (Table 13). The $[XeF_5]^+$ unit of $(XeF_6)_n$ is also preserved in the acceptor complexes as can be seen by comparison of the structural parameters of $[XeF_5]^+$ in the

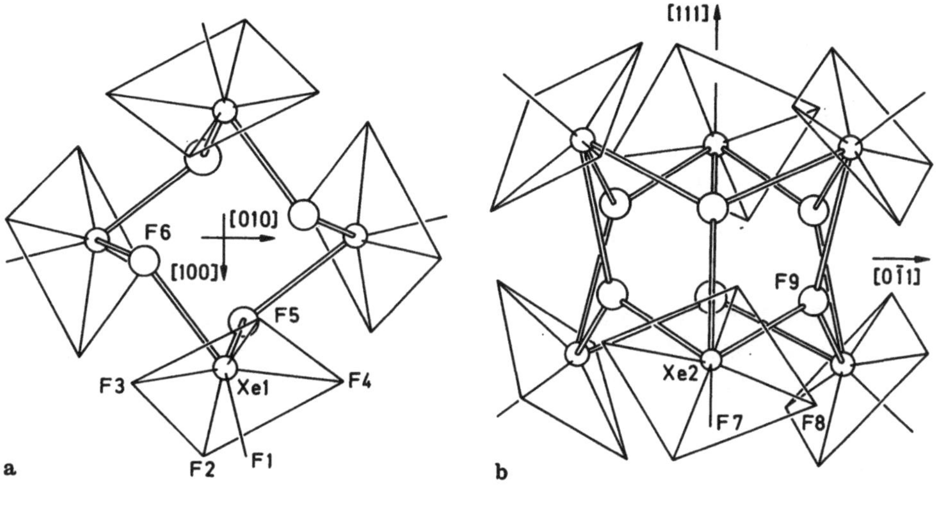

Fig. 6a and b. Molecular Geometries of the Cubic Phase of XeF_6

a Tetramer of $XeF_5^+F^-$ centered at $00^1/_4$ with the $\bar{4}$ axis parallel to [001]. Xenon atoms are indicated by small circles and bridging fluoride ions by large circles. XeF_5^+ ions are drawn in skeletal form to preserve clarity.

b Hexamer of $XeF_5^+F^-$ centered at $^1/_4{}^1/_4{}^1/_4$ and oriented with the 3 axis parallel to [111]. The three 2 axes are parallel to [0$\bar{1}$1], [10$\bar{1}$], and [$\bar{1}$10].

Table 13. Comparison of the Shapes of the $[XeF_5]^+$ Cation in XeF_6 Hexamers and Tetramers and in $[XeF_5]^+[AsF_6]^-$

	XeF_6 [a] Hexamers	XeF_6 [a] Tetramers	$[XeF_5]^+[AsF_6]^-$ [b]
$Xe-F_{ax}$, (Å)	1.76 (3)	1.84 (4)	1.78 (2)
$Xe-F_{eq}$, (Å)	1.92 (2)	1.86 (3)	1.83 (2)
$F_{ax}-F_{eq}$, (Å)	2.33 (3)	2.29	
$F_{eq}-F_{eq}$, (Å)	2.63 (3)	2.54 (13)	
$\measuredangle F_{ax}-Xe-F_{eq}$, (°)	80.0 (.6)	77.2 (1.8)	~80
$\measuredangle F_{eq}-Xe-F_{eq}$, (°)	88.3 (.2)	87.2	~90
$([F_5Xe]^+ \cdots F^-)$,	2.56	{ 2.23 2.60	
$\measuredangle([F_5Xe]^+ \cdots F^- \cdots [XeF_5]^+)$, (°)	118.8 (.3)	120.7 (1.2)	

[a] = Refs. [141, 142]; [b] = Ref. [37].

arsenic pentafluoride adduct, with the cation in the XeF_6 hexamer and tetramer (Table 13). It is interesting also that the bridging F^- ions are not close to the four-fold axis of the $[XeF_5]^+$ groups. Rather they are displaced by the lone pairs *trans* to the axial Xe—F bonds, thus preserving a pseudo-octahedral configuration.

b The structures of the Adducts in the Solid State and in Solution

The pseudo octahedral configuration and cationic nature of $[XeF_5]^+$ is thus amply confirmed in the solid phase. In fact, this unit has been established as present in a number of adducts through detailed single crystal structure determinations on; namely, $[XeF_5]^+[RuF_6]^-$ [34], $[XeF_5]^+[PtF_6]^-$ [117, 130] and $[XeF_5]_2^+[PdF_6]^{2-}$ [135], as well as less comprehensive data on $[XeF_5]^+[BF_4]^-$ [117], $[XeF_5]^+[AuF_6]^-$ [117] and $[XeF_5]^+[UF_6]^-$ [132]. The crystal data are given below and some of the molecular geometries are shown in Fig. 7.

$[XeF_5]^+[BF_4]^-$ [117] Orthorhombic, $a = 8.41$ (2), $b = 8.60$ (2), $c = 17.47$ (4) Å, $V = 1263.5$ Å^3, $Z = 8$. Space group unknown.

$[XeF_5]^+[RuF_6]^-$ [34] Orthorhombic, $a = 16.771$ (10), $b = 8.206$ (10), $c = 5.617$ (10) Å, $V = 773.03$ Å^3, $Z = 4$, $d = 3.79$ g cm^{-3}. Space group Pnma. $Xe-F_{ax} = 1.793$ (8) Å; $Xe-F_{eq} = 1.841$ (8) Å (twice); 1.848 (8) Å (twice); $\measuredangle F_{ax}-Xe-F_{eq} = 80°$.

$[XeF_5]^+[PtF_6]^-$ [170, 130] Orthorhombic, $a = 8.16$, $b = 16.81$, $c = 5.73$ Å, $V = 785.4$ Å^3, $Z = 4$. Space group Pmnb.
$Xe-F_{ax} = 1.81$ Å; $Xe-F_{eq} = 1.87$ Å (twice), 1.88 Å (twice); $\measuredangle F_{eq}-Xe-F_{ax} = 80°$.

$[XeF_5]^+[AuF_6]^-$ [117] Monoclinic, $a = 5.88$ (2), $b = 16.54$ (4), $c = 8.12$ (2) Å, $V = 791$ Å^3, $Z = 4$. Space group $P2_1/c$.

$[XeF_5]^+[UF_6]^-$ [132] Orthorhombic, $a = 9.39$ (2), $b = 19.86$ (4), $c = 8.41$ (2) Å, $Z = 8$, $d_c = 4.70$ g cm^{-3}. Space group Pcca or Pbcm.

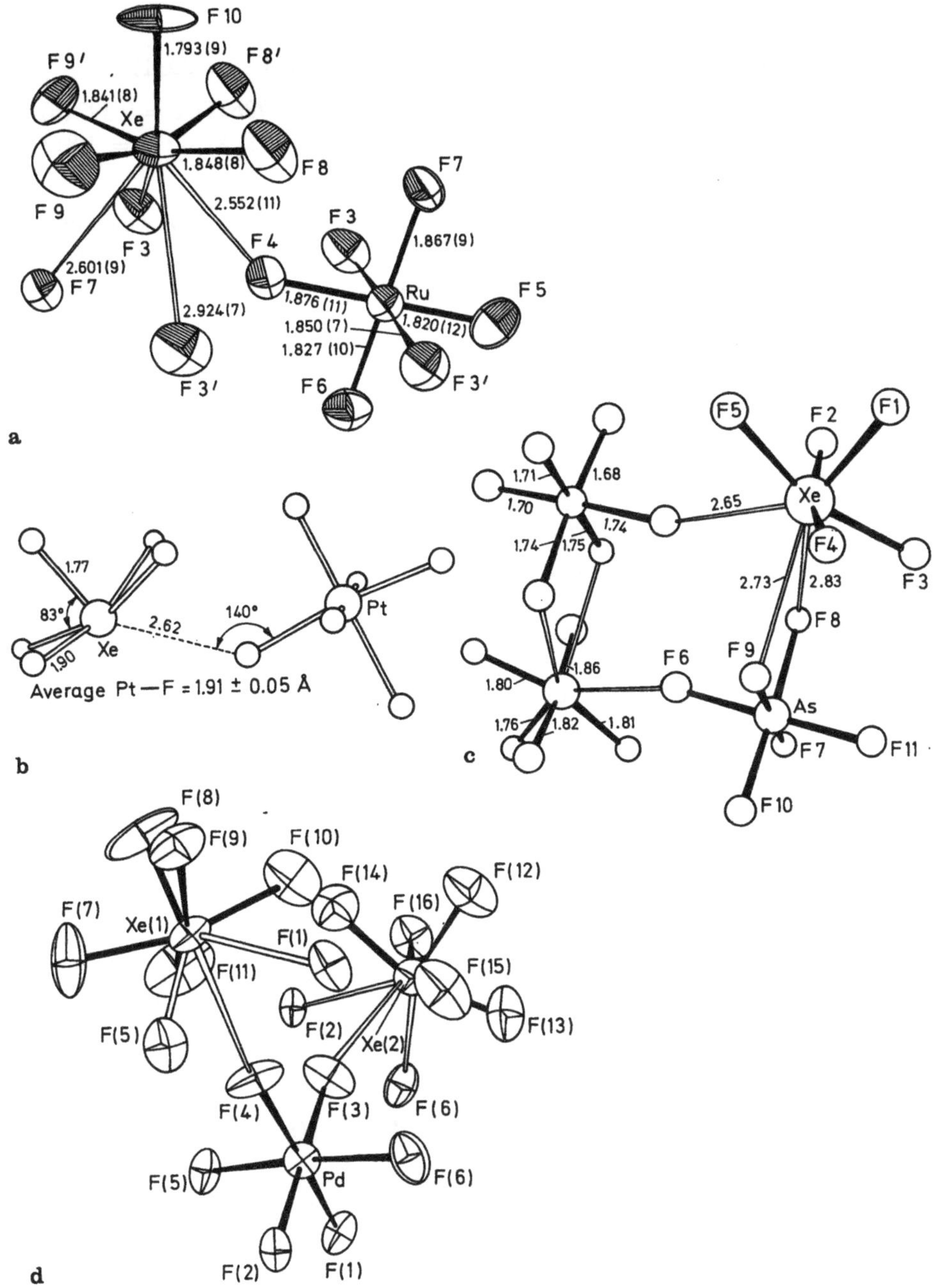

Fig. 7a—d. Molecular Geometries of Xenon Hexafluoride Adducts Containing the $[XeF_5]^+$ Cation. **a** The $[XeF^+][RuF_6^-]$ structural unit; **b** Configuration and bond distances for $[XeF_5]^+[PtF_6]^-$; **c** Configuration and bond distances for $[XeF_5^+][AsF_6^-]$; **d** The formula unit in $[XeF_5^+]_2[PdF_6^{2-}]$

$[XeF_5]^+[AsF_6]^-$ [37] Monoclinic, $a = 5.886\,(3)$, $b = 16.564\,(10)$, $c = 8.051\,(4)$ Å, $\beta = 91.57\,(3)°$, $V = 784.6$ Å^3, $Z = 4$, $d_c = 3.51$ g cm^{-3}. Space group $P2_1/c$. Xe—F_{ax} = 1.78 (2) Å, Xe—F_{eq} = 1.83 (2) Å, $\measuredangle$ F_{ax}—Xe—F_{eq} = 80 (1)°.

$[XeF_5]_2^+[PdF_6]^{2-}$ [135] Orthorhombic, $a = 9.346\,(6)$, $b = 12.786\,(7)$, $c = 9.397\,(6)$ Å, $V = 1122.9$ Å^3, $Z = 4$, $d_c = 3.91$ g cm^{-3}. Space group Pca2. Xe—F_{ax} = 1.813 Å, Xe—F_{eq} = 1.843 Å, $\measuredangle$ F_{eq}—Xe—F_{ax} = 79°.

All structures have several common features, the most prominent of which are the discrete XeF_5 groups of slightly distorted pseudo-octahedral symmetry as well as discrete, slightly distorted, MF_6 octahedra. Bond lengths and angles of the XeF_5 groups are rather similar.

The $[XeF_5]^+[AsF_6]^-$ and $[XeF_5]_2^+[PdF_6]^{2-}$ structures have the additional common feature that each cation makes *three* fluorine bridged contacts with two anions — i.e. association with one anion *via* a single fluorine bridge and with another anion *via* two, somewhat longer, fluorine bridges. These define $(XeF_5 \cdot MF_6)_2$ (M = As or Pd) centrosymmetric rings. The two sets of shorter and longer fluorine bridge contacts for $[XeF_5]^+[AsF_6]^-$ and $[XeF_5]^+[PdF_6]^{2-}$ are shown in Table 14.

The $[XeF_5]^+[RuF_6]^-$ and $[XeF_5]^+[PtF_6]^-$ structures have the common feature that each $[XeF_5]^+$ cation is coordinated to four; $[RuF_6]^-$ groups *via* one fluorine atom of the RuF_6 octahedra. The two sets of four Xe ··· F contacts are, for $[XeF_5]^+[RuF_6]^-$, 2.552 (11), 2.601 (9), 2.924 (7), 2.924 (7) Å, and for $[XeF_5]^+[PtF_6]^-$, 2.52, 2.65, 2.95, 2.95 Å.

In all four of the above structures the three (*viz.* four) incoming fluorine bridge atoms are arranged symmetrically around the pseudo-four fold axis of the XeF_5 group, approaching the latter from below the basal plane. This way the xenon valence lone pair is avoided and the fluorine bridge ligands approach the xenon in the direction which "sees" the highest density of positive charge. There is at present no adequate explanation why certain anions ($[RuF_6]^-$, $[PtF_6]^-$) provide four coordinating ligands while others ($[PdF_6]^{2-}$, $[AsF_6]^-$) provide only three ligands to the $[XeF_5]^+$.

The existence of $[XeF_5]^+$ as a discrete species in solution is also firmly established with the aid of ^{19}F [64,66,125] and ^{129}Xe [44,45] n.m.r. spectroscopy. The ^{19}F n.m.r. spectra of XeF_6 complexes of SbF_5, AsF_5 and BF_3 in a number of solvents such as SbF_5, BrF_5, HF and HSO_3F feature the AX_4 spectrum expected for a square pyramidal molecule. Xenon hexafluoride reacts with HSO_3F according to the equation (Eq. 24):

$$HSO_3F + XeF_6 \rightarrow [XeF_5]^+ + [SO_3F]^- + HF \qquad (24)$$

Table 14. Fluorine Bridge Contacts in Centrosymmetric Rings in $[XeF_5]^+[AsF_6]^-$ and $[XeF_5]_2^+[PdF_2]^{2-}$ Structures [37,135]

	Normal bridge bond lengths (Å)	Long bridge bond lengths (Å)
$[XeF_5]^+[AsF_6]^-$	2.65	2.73, 2.83
$[XeF_5]_2^+[PdF_6]^{2-}$	2.445	2.639, 2.559
	2.418	2.582, 2.617

The ^{19}F n.m.r. spectrum of this solution (—90 °C) shows the AX_4 spectrum, an HF line, and another line arising from exchange between $[SO_3F]^-$ and an excess of solvent. The spectrum of pure $[XeF_5]^+[SO_3F]^-$ dissolved in HSO_3F exhibits the same spectrum except for the line associated with HF [64, 123].

Representative n.m.r. parameters are given in Table 15. These examples are chosen to indicate the range of parameters. While ^{19}F chemical shifts and coupling constants are relatively constant, the ^{129}Xe chemical shifts and axial and equatorial coupling constants are very sensitive to solvent effects. This has been ascribed [64] to certain solvent interactions. The coupling constants, J_{Xe-F}, decrease with increasingly strong interactions with the solvent. By contrast, only the Xe—F equatorial coupling constants show significant temperature dependence.

An earlier empirical correlation [143] between ^{19}F chemical shifts and the ^{129}Xe—^{19}F coupling constants was extended to include the cations $[XeF_5]^+$, $[XeF_3]^+$, $[XeOF_3]^+$ and $[XeO_2F]^+$ [45]. This correlation provides a rationale for the small Xe—F couplings observed for the equatorial fluorines of $[XeF_5]^+$ if it is assumed that these change sign and, consequently, the absolute values of the axial coupling constants of $[XeOF_3]^+$ and $[XeF_5]^+$ are of opposite sign (i.e. negative) with respect to those of the other species.

Mössbauer spectra of $[XeF_5]^+[Sb_2F_{11}]^-$ [40] give no direct indication of the structure of $[XeF_5]^+$. They show, however, strong dependence of the electric field gradient on distortion of the octahedron. Thus, while the quadrupole splitting ΔE_Q, for neutral XeF_6 is 7.7 mm s^{-1}, for $[XeF_5]^+$ it is 12.0 mm s^{-1}. By contrast, little such dependence is seen in going from XeF_4 to $[XeF_3]^+$. This is probably because, in XeF_4 adducts, the square planar configuration around the xenon is essentially preserved, albeit distorted, but less so than in the $XeF_6 \rightarrow [XeF_5]^+$ transformation.

Initially, Raman spectra were used solely as a basis for fingerprinting and identifying the $[XeF_5]^+$ species [66, 83, 123, 132]. However, the infrared and Raman spectra of solid $[XeF_5]^+[BF_5]^-$ and $[XeF_5]^+[AsF_6]^-$ as well as their Raman spectra in HF solution were recorded [125], and Raman spectra of these salts as well as of several additional ones containing the $[AuF_6]^-$, $[PdF_6]^{2-}$, $[RuF_6]^{2-}$ and $[PtF_6]^-$ anions were also reported [117] (see Table 16).

Table 15. ^{19}F and ^{129}Xe Nuclear Magnetic Resonance Parameters for $[XeF_5]^+$

Solute	Solvent	δ^{19}F(ppm)[a]		δ^{129}Xe(ppm)[b]	Coupling Constants (Hz)		
		A	X_4		J_{FF}	$J_{^{129}Xe-^{19}F}$	
						A	X_4
$[XeF_5]^+[BF_4]^-$	HF (—80 °C)	—228.2	—106.2		182.0	1348	182.8
XeF_6	SbF_5 (35 °C)	—231.7	—108.8		175.7	1512	143.1
XeF_6	HSO_3F (—90 °C)	—226.2	—108.2		178.5	1357	175.0
$[XeF_5]^+[Sb_2F_{11}]^-$	HF (25 °C)			12.7		1400 (D)	159 (Q)
$[XeF_5]^+[SO_3F]^-$	HSO_3F (—80 °C)			—23.9		1377 (D)	165 (Q)

[a] With respect to external $CFCl_3$ (Refs. [64, 66, 125]); [b] With respect to external $XeOF_4$ (Refs. [44, 45]); D = doublet; Q = quintet.

Table 16. Fundamental Frequencies for $[XeF_5]^+$ (cm^{-1})

Assignment	Class	Symmetry Description	Adams, Bartlett (117)	Christe *et al.* (125)
ν_1	a_1	$\nu(XeF_{ax})$	670	679
ν_2		$\nu_s(XeF_4)$ in-phase	606	625
ν_3		$\delta_s(XeF_4)$ umbrella	312	355
ν_4	b_1	$\nu_{as}(XeF_4)$ out-of-phase)	600	610
ν_5		$\delta_{as}(XeF_4)$ out-of-phase	236	261
ν_6	b_2	$\delta_s(XeF_4)$ in-plane	[263]	300
ν_7	e	$\nu_{as}(XeF_4)$	677 } 644 }	652
ν_8		$\delta(F_{ax}XeF_4)$	416	410
ν_9		$\delta_{as}(XeF_4)$ in-plane	215	218

For $[XeF_5]^+$ of C_{4v} symmetry nine fundamental vibrations spanning the irreducible representations $3a_1 + 2b_1 + b_2 + 3e$ should be observed. All nine modes should be Raman active, but only the a_1 and e modes infrared active. On the basis of these spectra all nine fundamentals for $[XeF_5]^+$ were assigned and normal coordinate analysis carried out. The assignments for the fundamentals are given in Table 16.

Although both research groups made these assignments on the basis of comparison with the isoelectronic square pyramidal species IF_5, $[TeF_5]$ and $[SbF_5]^{2-}$, there are differences between the two sets. Discrepancies occur mainly with the fundamentals ν_3, ν_5, ν_6 and ν_7, the difficulties being due mainly to weaknesses of observed lines and masking by anion and cation lines. The earlier assignments [125] for ν_3 and ν_7 appear more reliable, because they are supported by relatively strong absorption bands in the infrared spectrum. The plausibilities of the assignments were examined by both groups by use of normal coordinate analysis. Because of the large number of force constants and small number of fundamentals, many simplifying assumptions had to be made involving transfer of interaction constants from related molecules and neglecting others. Good agreement between calculated and observed frequencies were obtained and sets of force constants were calculated which, in both cases, fit in well with trends established by the isoelectronic series $[SbF_5]^{2-}$, $[TeF_5]^-$, IF_5. It is thus impossible to choose between assignments on the basis of these calculations. The force constant trends illustrate well the decrease in bond polarity and increase in bond strengths and force constants in moving from left to right in the periodic system.

The pseudo-octahedral structure of the $[XeF_5]^+$ cation is also illustrated by the formation of the molecular adduct $XeF_2 \cdot 2([XeF_5]^+[AsF_6]^-)$ [83]. Crystals of the complex are monoclinic with a = 16.436 (10), b = 6.849 (6), c = 17.241 (11) Å, β = 93.2 (3)°, V = 1800 Å^3, Z = 4 and belong either to space group C2/c or Cc [83]. The XeF_2 molecules are probably surrounded symmetrically by $[XeF_5]^+$ ions in a way similar to that found in the $XeF_2 \cdot XeOF_4$ [83] and $XeF_2 \cdot IF_5$ [81] adducts, which contain the pseudooctahedral molecules $XeOF_4$ and IF_5.

3 $[Xe_2F_{11}]^+$

a Preparation of the Adducts

Compounds containing the $[Xe_2F_{11}]^+$ ion are prepared by fusing the fluoride acceptor molecule with an excess of XeF_6. The adduct $[Xe_2F_{11}]^+[AuF_6]^-$ is prepared by fluorination of AuF_3 in the presence of an excess of XeF_6. This salt reacts with CsF at 110 °C according to the Eq. (Eq. 25) [144]:

$$CsF + [Xe_2F_{11}]^+[AuF_6]^- \rightarrow CsAuF_6 + 2\,XeF_6 \qquad (25)$$

This method of replacing $[Xe_2F_{11}]^+$ by heavier alkali metal cations is probably quite general.

Of the adducts of XeF_6 listed earlier the following appear to contain the $[Xe_2F_{11}]^+$ cation:— $2\,XeF_6 \cdot RuF_5$, $2\,XeF_6 \cdot IrF_5$ [76], $2\,XeF_6 \cdot PtF_5$ [119], $2\,XeF_6 \cdot AuF_5$ [120, 122, 144], $2\,XeF_6 \cdot PF_5$ [117, 119, 134], $2\,XeF_6 \cdot AsF_5$ [76, 83, 117, 134], $2\,XeF_6 \cdot SbF_5$ [122] and $4\,XeF_6 \cdot PdF_4$ [120].

b The Structures of the Adducts in the Solid State and in Solution

The cationic nature of $[Xe_2F_{11}]^+$ in the solid state has been established by a crystal structure analysis of $[Xe_2F_{11}]^+[AuF_6]^-$ [120]. The crystals are orthorhombic with $a = 9.115\,(6)$, $b = 8.542$, $c = 15.726\,(20)$ Å, $V = 1224$ Å^3, $Z = 4$, $d_c = 4.24$ g $\cdot$ cm^{-3} and belong to space group Pnma. The $[Xe_2F_{11}]^+$ ion consists of two $[XeF_5]^+$ groups joined by a common fluorine atom (Fig. 8). The geometry of each XeF_5 group is quite similar to that of $[XeF_5]^+$ in $[XeF_5]^+[RuF_6]^-$ with very similar molecular parameters. The ionic description $[XeF_5]^+F^-[XeF_5]^+$ thus appears to provide a fairly good description of the species, although a certain measure of covalency in terms of $([XeF_5]^+XeF_6)$ should be introduced as suggested by departure from non-linearity of the Xe—F—Xe angle (169.2°). The interatomic bridging distances are 2.21 and 2.26 Å, respectively.

Raman spectra of $[Xe_2F_{11}]^+[MF_6]^-$ (M = P, As, Au) and $[Xe_2F_{11}]_2^+[PdF_6]^{2-}$ in the solid state and of $[Xe_2F_{11}]^+[MF_6]^-$ (M = P, As) in HF solution have been

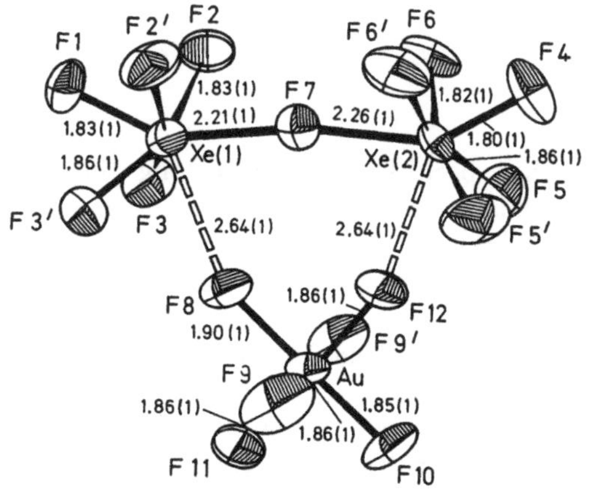

Fig. 8. Molecular Geometry of the $[Xe_2F_{11}]^+$ Cation

Table 17. Approximate Raman Frequencies and Partial Assignments for $[Xe_2F_{11}]^+$ (cm^{-1})[117]

Symmetry Description	Frequencies
$\nu(XeF_{ax})$	667 (p)
$\nu_{deg}(XeF_4)$	~640
$\nu_s(XeF_4)$	605 (p)
$\nu_{as}(XeF_4)$	~590 (dp)
$\delta(FXeF_4)$	410 (dp)
	360 (p)
$\delta_s(XeF_4)$	302 (p?)
$\delta_s(XeF_4)$ in-plane	~290
$\delta_{as}(XeF_4)$ out-of-plane	230
$\delta_{as}(XeF_4)$ in-plane	214 (dp)

(p) = polarized; (dp) = depolarized.

reported[117,120]. After anion lines are accounted for, a distinctive $[Xe_2F_{11}]^+$ spectral pattern emerges and partial assignments have been made (Table 17). Stretching frequencies of $[Xe_2F_{11}]^+$ are close to, but somewhat lower than, those of $[XeF_5]^+$ as might be expected from the lower overall charge per xenon atom. The only additional feature is the Raman line at 360 cm^{-1} (polarized in HF solution) which appears to be characteristic of the fluorine bridge betwen the XeF_5 groups.

The close analogies between the $([XeF_5]^+F^-)_n$ structure and those of the complexes containing $[XeF_5]^+$ and $[Xe_2F_{11}]^+$ are quite striking. Both cations seem to be derivative fragments of the "neutral" parent molecule, particularly the tetramer. This is also seen in Raman spectra of XeF_6 in HF as a function of concentration[117]. In concentrated solutions the pattern is reminiscent of $[XeF_5]^+F^-$, in intermediate concentrations of $[Xe_2F_{11}]^+$, and in dilute solutions of $[XeF_5]^+$. The principal equilibria can, therefore be described in terms of the equilibria outlined previously (see Sect. IV.1, Eqs. 22 and 23). The various salts isolated, therefore, are simple derivatives of the cationic species present in XeF_6 itself.

^{129}Xe Mössbauer spectra of $[Xe_2F_{11}]^+[BiF_6]^-$[40] show an isomer shift and quadrupole splitting characteristic of Xe(VI) complexes and are similar to those of $[XeF_5]^+[Sb_2F_{11}]^-$ and $[XeF_5]^+[SO_3F]^-$.

V $[XeOF_3]^+$

Although a complex, $XeOF_4 \cdot 2\,SbF_5$, has been known for a long time[145], recognition of its ionic character came much later when spectroscopic measurements began to be applied. Xenon oxide tetrafluoride is a relatively poor fluoride-ion donor and, consequently, only complexes with strong Lewis acids have been isolated. With SbF_5, only the 1:2 and 1:1 complexes are obtained depending on which reactant is used in excess. With AsF_5, an unstable adduct, $2\,XeOF_4 \cdot AsF_5$, has also been reported[146].

The compounds are prepared by mixing the components in appropriate ratios, either neat or in an HF solution, and pumping off the excess reactant to constant weight. $XeOF_4 \cdot SbF_5$ (m.p. 104–105 °C) [58,109] and $XeOF_4 \cdot 2\,SbF_5$ (m.p. 61–66 °C) [66,111,145] are both colourless solids.

No crystal structure determinations on the complexes have been carried out, but on the basis of n.m.r. [44,45,66,110] and Raman [58,66,110] spectroscopy the ionic nature of these complexes has been established. Fluorine-19 n.m.r. spectra in SbF_5 solutions show the existence of the cation, $[XeOF_3]^+$, and are consistent with a structure in which a lone pair, an oxygen atom and a fluorine occupy the equatorial positions and two fluorine atoms occupy the axial positions of a trigonal bipyramid [110]. Results from ^{129}Xe n.m.r. are also in accord with this interpretation [44,45]. Although the n.m.r. data (Table 18) could also be interpreted in terms of two equivalent equatorial fluorine atoms and one axial fluorine, Raman data [111] confirm the correctness of structure VIII, which is also consistent with VSEPRT.

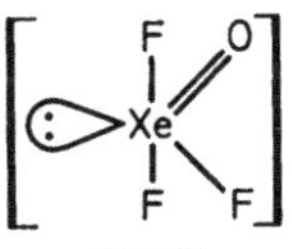

(VIII)

The assignment of cation and anion bands in the Raman spectrum was again complicated by overlap of the stretching fundamentals in the 450–750 cm^{-1} region and distortions of $[SbF_6]^-$ from octahedral symmetry, but there is good agreement in the results of the two studies that have been made [109,111], the earlier work [111] being more complete. The $[XeOF_3]^+$ cation is structurally related to the $[XeF_3]^+$ ion by replacing one of the equatorial lone pairs by an oxygen atom, and comparison of the spectra with those of $[XeF_3]^+$ and the isoelectronic IOF_3 and $ClOF_3$ molecules allowed complete vibrational assignments to be made. Assignments best fit structure VIII of C_s symmetry. The normal modes for this configuration span the irreducible representations $6a' + 3a''$, all of which are infrared and Raman active. The assignments are given in Table 19. Since many more bands are observed for the anion in $[XeOF_3]^+[SbF_6]^-$ than would be expected on the basis of O_h symmetry some fluorine-bridged interaction between cation and anion is implied and this is exemplified by a more satisfactory assignment on the basis of C_{4v} symmetry. Bands at 505 and 241 cm^{-1} were tentatively assigned to the $\nu_4(a_1)$ Sb—F stretching vibra-

Table 18. ^{19}F and ^{129}Xe Nuclear Magnetic Resonance Parameters for $[XeOF_3]^+$ and $[XeO_2F]^+$ [44,45,66,110]

		Chemical Shift (ppm)		Coupling Constants (Hz)		
		$^{19}F^a$	$^{129}Xe^b$	J_{FF}	$J_{^{129}Xe-^{19}F}$	
$[XeOF_3]^+$	A	−195.1 (T)	238	103.0	983	1018 (D)^c
	B_2	−147.1 (D)			434	434 (T)
$[XeO_2F]^+$		289.5	600		79.7	95 (D)

^{a} With respect to external $CFCl_3$; ^{b} With respect to external $XeOF_4$; ^{c} From ^{129}Xe n.m.r. spectrum.

Table 19. Raman Spectra of $[XeOF_3]^+$ Complexes [111]

$[XeOF_3]^+[SbF_6]^-$ (solid at −83 °C)	$[XeOF_3]^+[SbF_6]^-$ (~1 m HF soln)	$[XeOF_3]^+[Sb_2F_{11}]^-$ (solid at −83 °C)	$XeOF_4$ (0.8 m SbF_5 soln)	Assignments
44 (23)	942 p (100)	942 (70)	942 p, m	$\nu_1(a')$ $\nu(Xe{=}O)$
632 (32) 629 (34	637 p (59)	635 (90)	634 p, m	$\nu_2(a')$ $\nu(Xe{-}F_{eq})$
348 (3) 337 (7)		334 (11)		$\nu_3(a')$ $\delta_s(XeOF_{eq})$ in-plane bend
589 (41) 590 (100)	586 p (70)	610 (100) 598 (73)	589 p, m	$\nu_4(a')$ $\nu_s(Xe{-}F_{ax})$
206 (2)		199 (4)		$\nu_5(a')$ $\delta_s(XeOF_{eq})$ in-plane rock
194 (<1)				$\nu_6(a')$ $\delta_s(Xe{-}F_{ax})$
618 (15) 612 sh	612 (<1)	617 (15)		$\nu_7(a'')$ $\nu_{as}(XeF_{ax})$
366 (8)	365 (<1)	356 (14)		$\nu_8(a'')$ $\delta_{as}(XeOF_{eq})$
324 (9) 319 sh		319 (12)		$\nu_9(a'')$ $\tau(XeOF)$

tion of the Sb—F ··· Xe bridge and ν(Xe ··· F) vibration of the other half of the bridge respectively [111].

For $2\,XeOF_4 \cdot AsF_5$ no assignments have been given, but the appearance of a Raman band at 944 cm^{-1} in a solution of AsF_5 in $XeOF_4$ confirms the presence of $[XeOF_3]^+$ [146].

VI $[XeO_2F]^+$

XeO_2F_2 forms both 1:2 [66] and 1:1 [45] complexes with SbF_5. These are prepared by adding the appropriate amounts of SbF_5 to an HF solution containing XeO_2F_2 and removing excess of reactant and solvent by pumping.

The adduct $XeO_2F_2 \cdot SbF_5$ is a yellow solid which readily dissolves in SbF_5 at room temperature to give a yellow solution which slowly evolves a gas believed to be oxygen according to the equation (Eq. 26): —

$$[XeO_2F]^+ \xrightarrow[r.t]{SbF_5} [XeF]^+ + O_2 \qquad (26)$$

The appearance of $[XeF]^+$ as a function of time can be monitored by Raman spectroscopy [111]. Gas evolution ceases upon cooling to 5 °C. The ^{19}F n.m.r. spectrum of a solution of $[XeO_2F]^+[Sb_2F_{11}]^-$ in SbF_5 gives, in addition to the $[XeF]^+$ and F-on-Sb lines, an intense new singlet at low field which can be reasonably assigned to $[XeO_2F]^+$. The ^{19}F [66, 110] and ^{129}Xe [45] n.m.r. data are given in Table 18. The ^{129}Xe n.m.r. spectrum shows the presence of only one fluorine atom. An empirical plot of the ^{129}Xe chemical shifts vs. the number of Xe=O bonds for Xe(VI) cationic species ($[XeF_5]^+$, $[XeOF_3]^+$, $[XeO_2F]^+$) indicates the presence of two Xe=O bonds [45]. This correlation is attributed to monotonic deshielding of the central xenon atom with increasing oxygen substitution.

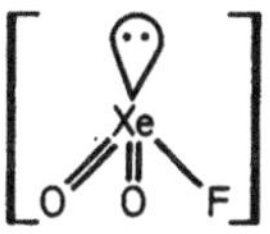

(IX)

From VSEPR theory it is expected that the $[XeO_2F]^+$ cation is a trigonal bipyramidal AX_3E molecule with C_s symmetry (structure IX). Six normal modes are expected for this structure, all of which would be Raman and infrared active. Partial assignments have been made on the basis of comparison with the related halogen oxide fluorides, XO_2F (X = Cl, Br, I). These are given in Table 20. No assignments were made for ν_3 and ν_6 which are expected to be weak and masked by the $[Sb_2F_{11}]^-$ bands. As in the case of $[XeF_3]^+/XeF_4$ and $[XeOF_3]^+/XeOF_4$ pairs, frequencies in $[XeO_2F]^+$ are higher than corresponding ones in the parent molecule, the increased charge on the xenon leading to stronger bonds in the cation.

Mössbauer spectra of the $[XeO_2F]^+[SbF_6]^-$ complex give an isomer shift of 0.1 (1) mm s^{-1} and a quadrupole splitting ΔE_Q = 13.6 (2) mm s^{-1}. The latter low value is characteristic of Xe(VI) compounds.

Table 20. Observed Raman Frequencies (cm^{-1}) and Assignments for $[XeO_2F]^+$ in $[XeO_2F]^+[Sb_2F_{11}]^-$ [66,111]

$[XeO_2F]^{+\,a}$	Assignment
923 (38)	$\nu_5(a'')$ $\nu(XeO_2)$ asym. str.
867 (100)	$\nu_1(a')$ $\nu(XeO_2)$ sym. str.
580 (58)	$\nu_2(a')$ ν(Xe—F)
334 (23)	$\nu_4(a')$ $\delta(XeO_2)$ sym. bend
	$\nu_3(a')$ (Xe—F) in-plane def.
	$\nu_6(a'')$ (Xe—F) out-of-plane def.

[a] The spectrum was recorded at —107 °C.

VII $[XeOF_5]^+$

Although a complex of composition $[XeOF_5]^+[Sb_nF_{5n+1}]^-$ has been reported [58] to result from the oxidation of $XeOF_4$ by $[KrF]^+$, the actual reaction product has been shown to be a complex containing the $(XeOF_4 \cdot [XeF_5]^+)$ cation [11] (see Sect. II, 3, b).

VIII $[XeOTeF_5]^+$

Xenon difluoride reacts with pentafluoroorthotelluric acid to give xenon(II) fluoride orthopentafluorotellurate. The latter reacts with AsF_5 according to the equation (Eq. 27): —

$$FXeOTeF_5 + AsF_5 \rightarrow [XeOTeF_5]^+[AsF_6]^- \quad (27)$$

This new adduct is a bright yellow solid melting at 160 °C [147,148]. The Raman spectrum shows lines characteristic of hexafluoroarsenate, and the absence of an Xe—F vibration justifies its formulation as an ionic salt. The parent, $FXeOTeF_5$, is a poorer donor than XeF_2 since it is formed by reaction of the salt with XeF_2 (Eq. 28).

$$[XeOTeF_5]^+[AsF_6]^- + 2\,XeF_2 \rightarrow [Xe_2F_3]^+[AsF_6]^- + FXeOTeF_5 \quad (28)$$

Dissolution of $[XeOTeF_5]^+[AsF_6]^-$ in SbF_5 results in displacement of AsF_5 by SbF_5 and formation of a yellow orange solution (Eq. 29) [149].

$$[XeOTeF_5]^+[AsF_6]^- + nSbF_5 \rightarrow [XeOTeF_5]^+[Sb_nF_{5n+1}]^- + AsF_5 \quad (29)$$

^{19}F, ^{129}Xe and ^{125}Te n.m.r. spectra in SbF_5 solution have confirmed the discrete nature of the $[XeOTeF_5]^+$ ion showing *a*) in the ^{19}F n.m.r., the AB_4 ($J_{F_{ax}-F_{eq}}/\nu_0\delta_{F_{ax}-F_{eq}} = 0.1497$) spectrum of two sets of fluorine atoms on Te, *b*) in the ^{125}Te—F and ^{125}Te—F_b (Formula X) and *c*) in the ^{129}Xe n.m.r., long range ^{129}Xe—$^{19}F_b$ coupling (Formula X). The n.m.r. parameters are shown in Table 21.

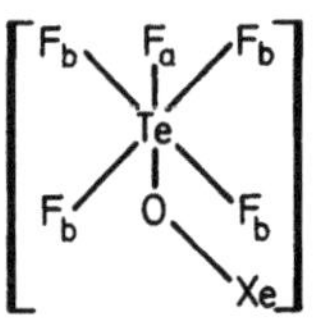

(X)

Raman spectra were assigned on the basis of C_s symmetry with a staggered conformation [149]. A total of 18 Raman and infrared active modes are predicted for this configuration with symmetries 11A′ + 7A″. Assignments, made by comparison with the related TeF_5Cl molecule of C_{4v} symmetry, are given in Table 22. The

Table 21. N.m.r. Parameters for $[XeOTeF_5]^+[AsF_6]^-$ in SbF_5 at 25 °C [149]

Chemical Shifts (ppm)			Spin-spin Coupling Constants (Hz)		
$\delta_{^{19}F}$	$\delta_{^{129}Xe}$	$\delta_{^{125}Te}$	$J_{F-F'}$	$J_{^{129}Xe-^{19}F}$	$J_{^{125}Te-^{19}F}$
F, −41.0 F′, −54.6	−1472	−134.9	172.2 (AB_4)	F, 18.5 F′, not resolved	F, 3814 (AB_4X) F′, 3802 (AB_4X)

Table 22. Observed Raman Frequencies (cm^{-1}) and Assignments for $[XeOTeF_5]^+$ [149)]

$[XeOTeF_5]^+[Sb_2F_{11}]^-$ [a]	$[XeOTeF_5]^+[AsF_6]^-$ [a]	Assignment Description
748 (2)	739 (6)	a″, $\nu_{as}(TeF_4)$, asym to plane of symmetry
741 (14)	775 (20)	a′, $\nu_{as}(TeF_4)$, sym to plane of symmetry
714 (23)	713 (34)	a′, $\nu(TeF')$
671 (64)	668 (100)	a′, $\nu_s(TeF_4)$ breathing
661 (31)	663 (58)	a″, $\nu_s(TeF_4)$ out-of-phase
487 (41)	{ 492 (16) / 483 (14)	a′, $\nu_s(XeOTe)$
478 sh	{ 470 (18) / 476 sh	a′, $\nu_{as}(XeOTe)$
	333 (20	a″, $\delta(F'TeF_4)$, out-of-plane of symmetry
320 (4)	320 (7)	a′, $\delta(F'TeF_4)$, in plane of symmetry
311 (10)	312 (8)	a′, $\delta_s(TeF_4)$, out-of-plane
293 (9)	295 (3)	a′, $\delta(TeF_4)$, in-plane scissors
		a′, $\delta_{as}(TeF_4)$, sym to plane of symmetry
252 (28)	252 (25)	
		a″, $\delta_{as}(TeF_4)$, asym to plane of symmetry
184 (4)	191 (5)	a′, $\delta(OTeF_4)$ in-plane
210 (3)	205 (1)	a″, $\delta(OTeF_4)$ out-of-plane
		a″, $\delta_{as}(TeF_4)$ out-of-plane
173 (31)	174 (32)	a′, $\delta(XeOTe)$
		a″, $\tau(Xe—O—TeF_5)$
	365 (15)	$\nu(Xe \cdots F)$

[a] Spectra recorded at —196 °C.

Raman lines assigned to the $[AsF_6]^-$ anions evidence considerable departure from octahedral symmetry indicating significant Xe ··· F—As bridging interaction.

^{129}Xe Mössbauer parameters are $\Delta E_Q = 37.4$ (1) mm s^{-1} and $S = 0.17$ (5) mm s^{-1} [40)]. These indicate reduction of charge transfer which is suggestive of O—Xe—F chains as observed also in the Raman spectra.

The $[XeOTeF_5]^+$ cation undergoes some interesting reactions during the course of which several, as yet not fully characterized, xenon cations are formed [149)].

Dissolution of $[XeOTeF_5]^+[AsF_6]^-$ in BrF_5 at —48 °C yields a bright yellow solution which disappears upon warming briefly to room temperature. Low-temperature ^{129}Xe and ^{19}F both before and after warming are consistent with the following reactions (Eqs. 30–32)

$$2\,[XeOTeF_5]^+[AsF_6]^- + BrF_5 \rightarrow TeF_6 + \underset{(XI)}{[FXeFXeOTeF_5]^+[AsF_6]^-} + [BrOF_2]^+[AsF_6]^- \quad (30)$$

$$[FXeFXeOTeF_5]^+[AsF_6]^- + BrF_5 \rightarrow TeF_6 + [Xe_2F_3]^+[AsF_6]^- + BrOF_3 \quad (31)$$

$$[Xe_2F_3]^+[AsF_6]^- + BrOF_3 + [BrOF_2]^+[AsF_6]^- \rightarrow \rightarrow 2\,(XeF_2 \cdot [BrOF_2]^+[AsF_6]^-) \quad (32)$$
(XII)

Species (XI) (Eq. 30), which is stable at low temperature, is thought to be responsible for the yellow colour. The disappearance of the colour on warming yields n.m.r. results consistent with Eq. 31. A final reaction (Eq. 32) is believed to occur rapidly at room temperature. The solution behaviour of the complex (Formula XII) and Raman spectra of the solid, which can be isolated by pumping off the solvent at —48 °C, suggest association through covalent bonding between XeF_2 and $[BrOF_2]^+$. Spectral assignments to the $XeF_2 \cdot [BrOF_2]^+$ cation have been made on the basis of C_1 symmetry for which 15 A-type infrared and Raman active modes are expected [149].

Evidence for another new cation, $[XeOSO_2F]^+$, has been found by reaction of $[XeOTeF_5]^+[AsF_6]^-$ in HSO_3F at low temperature [149]. Previous attempts to prepare this species by interaction of $FXeSO_3F$ with AsF_5 at —78 °C were unsuccessful [150].

The appearance of $HOTeF_5$ and $[OTeF_5]^+$ in such solutions at —10 °C have been explained by the equilibrium (Eq. 33):

$$HOSO_2F + [XeOTeF_5]^+ \rightleftharpoons HOTeF_5 + [XeOSO_2F]^+ \quad (33)$$
(XIII)

A new peak in the ^{129}Xe n.m.r. spectrum at about —1290 ppm is believed to arise from species (XIII) (Eq. 33) which, upon warming decomposes according to the equation (Eq. 34).

$$2\,([XeOSO_2F]^+[AsF_6]^-) \rightarrow Xe + [XeF]^+[AsF_6]^- + AsF_5 + S_2O_6F_2 \quad (34)$$

IX $[(FXeO)_2S(O)F]^+$

Many complexes of xenon(II) have been prepared by substituting one or both fluorine atoms in XeF_2 with highly electronegative ligands or by fluorine transfer to strong Lewis acids. Some of the resulting complexes have served as starting materials for formation of the novel ion $[(FXeO)_2S(O)F]^+$. Salts containing this ion have been prepared by the following reactions (Eqs. 35–37) [150, 151]:

$$2XeF_2 + HSO_3F + AsF_5 \xrightarrow[-78\,°C]{HF} [(FXeO)_2S(O)F]^+[AsF_6]^- + HF \quad (35)$$

$$[Xe_2F_3]^+ + HSO_3F \xrightarrow[\text{low temp.}]{HF} [(FXeO)_2S(O)F]^+ + HF\,. \quad (36)$$

$$2([XeF]^+[AsF_6]^-) + KSO_3F \xrightarrow[-78\,°C]{HSO_3F} [(FXeO)_2S(O)F]^+[AsF_6]^- + KAsF_6 \quad (37)$$

The first two reactions differ only in so far as whether the $[Xe_2F_3]^+$ is prepared first or *in situ*. The salt is a colourless solid which sublimes at ~20 °C in vacuo. It decomposes above 64 °C according to the equation (Eq. 38):

$$4\,[(FXeO)_2S(O)F]^+[AsF_6]^- \rightarrow 3\,[Xe_2F_3]^+[AsF_6]^- + AsF_5 + 2\,Xe + 2\,S_2O_6F_2 \qquad (38)$$

The cation is unstable in an excess of HSO_3F at room temperature, slowly decomposing according to the equation (Eq. 39):

$$[(FXeO)_2S(O)F]^+ + HSO_3F \rightarrow [XeF]^+ + Xe + S_2O_6F_2 + HF \qquad (39)$$

^{19}F n.m.r. spectra give three lines of relative intensites 1:6:2 corresponding to F-on-S, F-on-Xe, and compatible with the structure shown (XIV). N.m.r. parameters

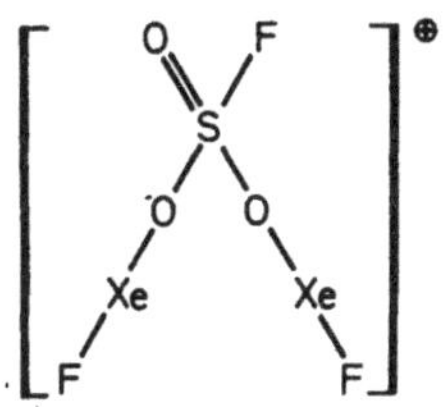

(XIV)

are given in Table 23. The fluorosulphate bridged structure of the cation has been confirmed by a crystal structure determination, the crystal data for which are presented below:

$[(FXeO)_2S(O)F]^+[AsF_6]^-$ [152]. Monoclinic, $a = 11.178$ (5), $b = 8.718$ (5), $c = 11.687$ (6) Å, $\beta = 91.28$ (4)°, $V = 1132$ Å^3, $Z = 4$, $d_c = 3.45$ g cm^{-3}. Space group $P2_1/n$.

The structural unit consists of $[(FXeO)_2S(O)F]^+$ cations and $[AsF_6]^-$ anions with no cation-anion contacts of less than 3.0 Å. The cation consists of two XeF groups bonded to an approximately tetrahedral SF_3O group. The S—F bond length is

Table 23. N.m.r. Parameters for $[(FXeO)_2S(O)F]^+$

Chemical Shifts (ppm)					Coupling Constants	Ref.
Solute	Solvent	$\delta_{F(ppm)}$		$\delta_{Xe(ppm)}$	(Hz) J_{Xe-F}	
$[Xe_2F_3]^+[AsF_6]^-$ (−91°)	HSO_3F	−44.7	~221.0		~6380	[151]
$[(FXeO)_2S(O)F]^+[AsF_6]^-$ (−59°)	BrF_5	−44.4	221.9		6470	
$[(FXeO)_2S(O)F]^+[AsF_6]^-$ (−77°)	BrF_5			−1258(D)	6428	[45]

1.53 Å and the unique S—O bond length 1.39 Å. The two other S—O bond distances are 1.47 Å. The Xe—F distances are 1.86 Å and the Xe—O distances ~2.21 Å. These are compatible with a bonding description of the cation in terms of contributions from the structures (XV–XVII).

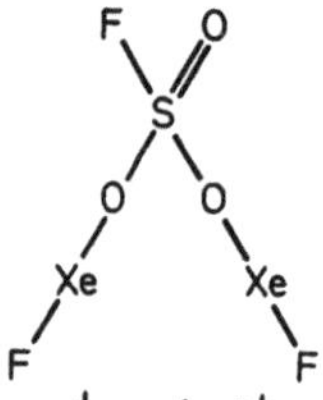
'covalent'
(XV)

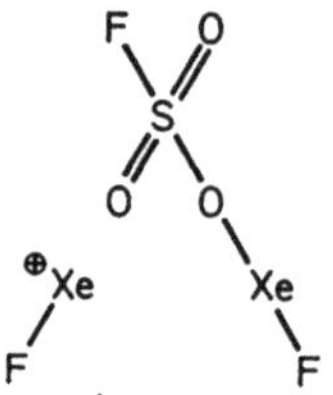
'ionic'
(XVI)

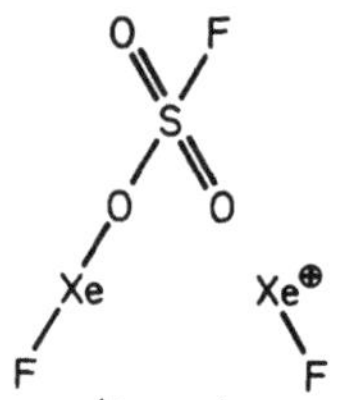
'ionic'
(XVII)

Raman spectra have been reported by two groups[150,151]. Agreement is good but the crystal structure has shown that the cation has C_1 symmetry rather than C_s on the basis of which the earlier Raman assignments had been made. Only the more complete data are presented in the Table (Table 24). In the C_1 configuration 21 normal modes of symmetry a, all infrared and Raman active, are expected for $[(FXeO)_2S(O)F]^+$. Only lines for a symmetrical octahedral $[AsF_6]^-$ anion are observed, indicating the absence of fluorine bridges and in accord with the absence of interionic contacts of less than 3 Å. The structure is thus purely ionic in character.

^{129}Xe Mössbauer spectra give an isomer shift of 0.1 (2) mm s^{-1} and a quadrupole splitting ΔE_q = 40.5 (4) mm s^{-1}, the latter being somewhat smaller than for species

Table 24. Observed Raman Frequencies (cm^{-1}) and Assignments for $[(FXeO)_2S(O)F]^+$ [151]

$[(FXeO)_2S(O)F]^+$	Assignment
1342 (8)	S=O str.
1084 (2)	SO_2, asym str.
1032 (8)	SO_2, sym. str.
871 (3)	S—F str.
629 (15)	S=O wag
594 (8)	SO_2 bend
582 (52)	Xe—F asym str.; Xe—F sym str.
571 (100	
566 (36)	
554 (82)	
540 (3)	S—F wag
400 (6)	Xe—O asym str.; Xe—O sym str.
395 (3)	
392 (20	SO_2 rock
255 (3)	SO_2F torsion
245 (1)	O—Xe—F sym and asym, in-plane and out-of-plane bends
240 (1)	
193 (4)	Xe—O—S sym and asym, in-plane and out-of-plane bends
112 (4)	
138 (3)	

with F ··· Xe—F chains. This suggests smaller charge transfer in the Xe—O than the Xe—F bond.

X $[(FO_2S)_2NXe]_2F^+$

This ion is surmised to exist in the product formed by the reaction of $FXeN(SO_2F)_2$ with AsF_5 according to the reaction sequence (Eqs. 40, 41) [153]:

$$FXeN(SO_2F)_2 + AsF_5 \xrightarrow[16\,h]{-125\text{ to }-10\,°C} \underset{(XVIII)}{FXeN(SO_2F)_2 \cdot AsF_5} + AsF_5 + Xe \quad (40)$$

$$FXeN(SO_2F)_2 \xrightarrow[4h,\ vac]{22\,°C} 2FXeN(SO_2F)_2 \cdot AsF_5 + AsF_5 + Xe(trace). \quad (41)$$

The 1:1 adduct (XVIII), a bright yellow solid, is unstable at room temperature while the pale yellow solid 2:1 adduct appears to be stable. Raman spectra show complete absence of ν(Xe—F) at 504 cm^{-1} indicating the absence of terminal xenon-fluorine bonds. The $\nu_{sym}(SO)$ and $\nu_{asym}(SO)$ bands are shifted to 1236 and 1494 cm^{-1}, which are higher than those in the neutral molecule, suggesting cation formation.

No signal was obtained in the ^{19}F n.m.r. spectrum for the bridging fluorine, but this could be due to exchange broadening.

C Anions

I $[XeF_8]^{2-}$ and $[XeF_7]^-$

Complexes of XeF_6 with fluorine ion donors such as NOF, NO_2F and alkali metal fluorides are well known. With alkali fluorides complexes of the type $MXeF_7$ (M = Cs, Rb) and M_2XeF_8 (M = Cs, Rb, K, Na) have been prepared [154,155], but although they probably contain the anions $[XeF_7]^-$ and $[XeF_8]^{2-}$, there is thus far no structural evidence to support this. On the other hand 2 NOF · XeF_6, the product of the reaction of XeF_6 with NOF, has been fully characterized. The white solid, which sublimes at room temperature exhibits characteristic vibrations of $[NO]^+$ and ν(Xe—F) in its infrared and Raman spectra, which suggested the formulation $[NO]_2^+[XeF_8]^{2-}$ [156]. Firm evidence for the existence of $[XeF_8]^{2-}$ was obtained from a crystal structure determination (Fig. 9). The crystal data are summarized below.

$[NO]_2^+[XeF_8]^{2-}$ [157] Orthorhombic, $a = 8.914(10)$, $b = 5.945(10)$, $c = 12.83(2)$ Å, $V = 679.9$ Å^3, $Z = 4$, $d_c = 3.354$ g cm^{-3}, Space group Pnma.

The anion can be described as a slightly distorted Archimedean antiprism with Xe—F distances ranging from 1.946 to 2.099 Å. The eight-fold xenon coordination provides no clearly defined position for the lone pair, thus seemingly being in violation of the VSEPR theory.

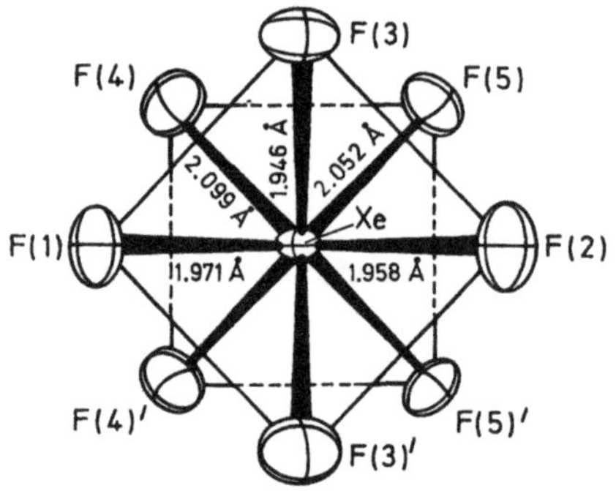

Fig. 9. Molecular Geometry of the $[XeF_8]^{2-}$ Anion

Crystal field calculations on the ground state energy of $[NO]_2^+[XeF_8]^{2-}$ confirm that the minimum energy occurs for a D_{4d} structure with $\theta = 57°$ (the average polar angle for the ligands with respect to an approximate S_8 axis). However, geometries which would permit appreciable 5s-5p mixing, and thus allow for the lone pair, were not taken into account [158].

No 1:1 complex of NOF with XeF_6 is known. However, $NO_2F \cdot XeF_6$ has been isolated [159]. Although infrared spectra of the solid at —195 °C show absorptions characteristic of $[NO_2]^+$ cations, which would imply the presence of $[XeF_7]^-$, relatively small shifts of ν(Xe—F) suggest rather strong fluorine bridging and that the formulation $[NO_2]^+[XeF_7]^-$ is probably not justified [159].

II $[XeOF_5]^-$ and $[Xe_3O_3F_{13}]^-$

Complexes of $XeOF_4$ with fluoride ion donors such as CsF, RbF, KF [145] and NOF [156] have been known for some time. The former are prepared by treating the alkali fluorides with an excess of $XeOF_4$ and pumping to constant weight at room temperature. In the case of CsF, the compound $CsF \cdot 3\,XeOF_4$ has also been isolated at 0 °C. This salt decomposes to give the 1:1 adducts under dynamic vacuum [160]. All of the adducts decompose at higher temperatures giving a number of intermediate stoicheiometries as summarized in Eqs. 42–44.

$$CsF \cdot 3XeOF_4 \xrightarrow[-XeOF_4]{20\,°C\ vac} CsF \cdot XeOF_4 \xrightarrow[-XeOF_4]{\sim 125\,°C} 3CsF \cdot 2XeOF_4 \xrightarrow[-XeOF_4]{\sim 270\,°C} 3CsF \cdot XeOF_4 \xrightarrow[-XeOF_4]{\sim 400\,°C} CsF \quad (42)$$

$$3RbF \cdot 2XeOF_4 \rightarrow \text{intermediate stoicheiometries of unknown composition} \xrightarrow[-XeOF_4]{\sim 400\,°C} RbF \quad (43)$$

$$3KF \cdot XeOF_4 \xrightarrow[-XeOF_4]{\sim 90\,°C} 6KF \cdot XeOF_4 \xrightarrow[-XeOF_4]{\sim 250\,°C} KF\,. \quad (44)$$

A shift in the Xe=O stretching frequency from 923 cm^{-1} in $XeOF_4$ to ~880 cm^{-1} in solid alkali fluoride-xenon oxide tetrafluoride adducts has been taken to indicate formation of $[XeOF_5]^-$ [146]. Raman spectra of NOF · $XeOF_4$ show that no appreciable fluoride ion transfer occurs [156].

Firm identification of the $[XeOF_5]^-$ ion in the solid state for the alkali fluoride adducts was based on more detailed analysis of the Raman spectra [160]. VSEPR theory predicts distortion from the pseudo octahedral (C_{4v}) to C_s symmetry, the lone pair occupying an octahedral face adjacent to the axial fluorine (XVIII). For

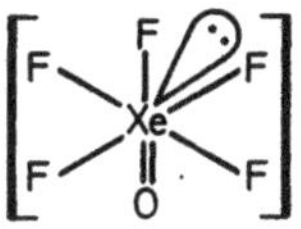

(XVIII)

this structure 15 Raman active bands, 9a′ + 6a″, are expected. Assignments, based on the C_{4v} structure and comparison with the related molecules, IOF_5 and $[OTeF_5]^-$, are given in Table 25. The e modes are split into a′ and a″ components under C_s symmetry, but more definite assignments are not possible on the basis of Raman spectra alone.

Raman spectra on CsF · 3 $XeOF_4$ have shown the presence of the $[Xe_3O_3F_{13}]^-$ ion [160]. This consists of three $XeOF_4$ molecules bonded equivalently to a central F^- ion. The expected point symmetry is C_{3v}. The proposed formulation is supported by ^{16}O-^{18}O isotope substitution experiments. X-ray powder photography showed [160] that the compound is isostructural with CsF · 3 IF_5 [161], but definitive confirmation of the structure has had to wait for a single crystal study, the data from which are outlined below (Fig. 10).

Table 25. Observed Raman Frequencies and Assignments for $[XeOF_5]^-$ [160]

$[XeOF_5]^-$	Assignment Description
883 (66)	a_1, ν(Xe=O)
544 sh	b_1, $ν_s(XeF_4)$ out-of-plane
524 (100)	a_1, $ν_s(XeF_4)$
473 (34)	
468 (29)	e, $ν_{as}(XeF_4)$
435 (12)	
420 (10)	a_1, ν(Xe—F′)
410 (7)	e, δ(F′XeF_4), (F′XeO) bend
396 (24)	
390 (40)	b_2, $δ_s(XeF_4)$ out-of-plane
361 (17)	a_1, $δ_s(XeF_4)$ in-plane
384 (18)	e, δ($OXeF_4$), (OF_3) scissor
365 (15)	
293 (13)	e, $δ_{as}(XeF_4)$ in-plane puckering
274 (8)	
177 (1)	b_1, $δ_{as}(XeF_4)$ out-of-plane

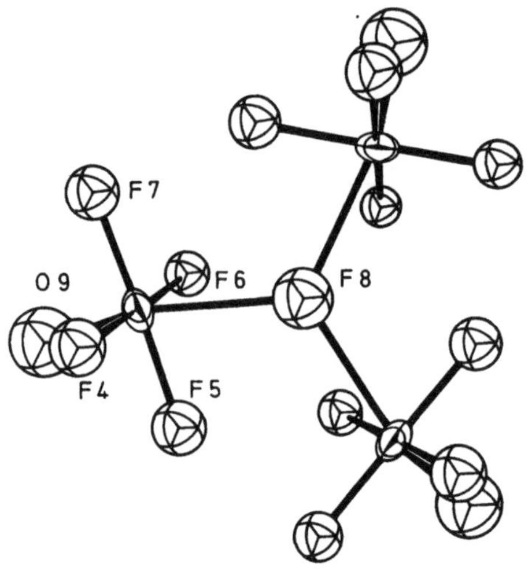

Fig. 10. Molecular Geometry of the $[(XeOF_4)_3F]^-$ Anion

$[(XeOF_4)_3F]^-$ Cs^+ [162] Cubic, a = 13.933 (7) Å, V = 270.8 Å^3, Z = 4, d_c = 4.035 g cm^{-3}. Space group Pa3.

In the anion structure three $XeOF_4$ units, which have close to a square pyramidal structure, are linked via three equal Xe ··· F bonds to a common fluorine, the arrangement being similar to that of the hexameric $(XeF_6)_6$ units in the cubic phase of XeF_6.

III $[XeO_3X]^-$ (X = F, Cl, Br)

Hydrolysis of $CsXeOF_5$ or $CsXeF_7$ leads to the same compound of stoicheiometry $CsF \cdot XeO_3$ [145]. Because of an absence of infrared absorptions in the Xe—F bond-stretching region, this was originally thought to be a molecular complex. A subsequent single crystal structure analysis of $KF \cdot XeO_3$ showed that this class of compounds should be formulated as $nM^+[XeO_3F]_n^-$ (M = Cs, Rb, K) [163].

More convenient methods of preparation consist of mixing appropriate amounts of solutions of aqueous XeO_3 and MF and subsequent evaporation until crystals are obtained [164]. The caesium and rubidium salts have also been prepared by neutralizing an aqueous XeO_3 solution containing HF from the hydrolysis of XeF_6 with appropriate amounts of CsOH or RbOH, and evaporating to dryness. The haloxenates, $CsXeO_3Cl$ [165] and $CsXeO_3Br$ [164] have been obtained by similar methods.

The thermal stabilities of these salts decrease with increasing atomic weight of the halogen, $CsXeO_3F > CsXeO_3Cl > CsXeO_3Br$. The fluoride salts are stable to 200 °C, but decompose to xenon, oxygen and alkali fluoride above 260 °C. The chloride xenates begin to decompose at 150 °C (and explosively so above 200 °C) [165]. The bromo salt is quite unstable even at room temperature.

Attempts to prepare $NaXeO_3F$ yielded products contaminated with NaF or XeO_3 due to the relatively high solubility of this fluoroxenate salt.

Crystal structure data for $KXeO_3F$ are summarized below.

$[XeO_3F]^-K^+$ [163] Orthorhombic, $a = 7.374$ (5), $b = 6.811$ (5), $c = 8.185$ (6) Å, $V = 411.1$ Å^3, $Z = 4$, $d_c = 3.835$ g cm^{-3}. Space group $Pn2_1a$.

The structure consists of infinite chains of XeO_3 units linked by fluorine bridges with potassium atoms at non-bridging distances (Fig. 11). The geometry of XeO_3 in the anion is similar to that of XeO_3 itself with Xe—O bond lengths of 1.75 (1), 1.76 (1) and 1.79 (1) Å and O—Xe—O bond angles of 97.8 (7), 100.5 (1.2) and 101.1 (9)°. The two bridging Xe—F distances are 2.36 (1) and 2.48 (1) Å, considerably longer than those of non-bridging Xe—F bonds (~1.9 Å), but significantly shorter than non-bonding xenon-fluorine interactions (~3.5 Å). The compound can thus be formulated as $nK^+[XeO_3F]_n^-$.

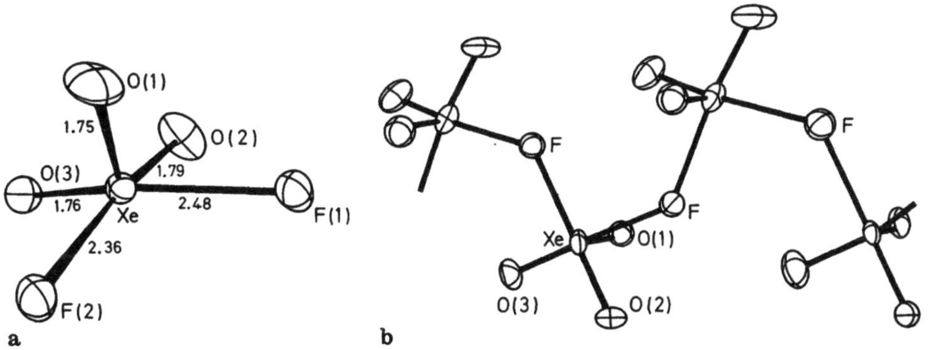

Fig. 11a and b. Molecular Geometry of the $[XeO_3F]^-$ Anion. **a** The inner coordination geometry around XeO_3; $KXeO_3F$. **b** A view of the XeO_3F^- anion showing the bridge Xe—F—Xe bonds and the polymeric nature of the ion

The existence of $[XeO_3Cl]^-$ in the solid state has been verified similarly by neutron and X-ray diffraction analysis of the compounds 2.25 MCl · XeO_3 (M = Cs, Rb) [166]. The crystals are body-centred tetragonal with $a = 16.56$, $c = 7.10$ Å and $a = 15.97$, $c = 6.91$ Å for the caesium and rubidium salts respectively, and contain eight formula units per unit cell. The structures were refined in space group $I\bar{4}$. The structures feature infinite chains of $[XeO_3Cl]^-$ units linked by nearly linear chlorine bridges into $[XeO_3Cl_2)_n^{2n-}$ chains. The xenon coordination consists of a distorted octahedral XeO_3Cl_3 moiety. The Xe—O bond lengths are similar to those of XeO_3 itself. The single terminal and two bridging Xe—Cl bond lengths are nearly equivalent at ~2.95 Å, considerably shorter than the van der Waals sum of 4.02 Å, but about 0.5 Å longer than the expected Xe—Cl single bond length. This is analogous to the lengthening of Xe—F in $[XeO_3F]^-$.

Small amounts of a second, triclinic, phase have been detected in both the caesium and rubidium salts. For the rubidium compound cell parameters are $a = 10.10$, $b = 12.77$, $c = 10.00$ Å, $\alpha = 91.30$, $\beta = 92.2$ and $\gamma = 100.64°$.

Although the site symmetry of the $[XeO_3F]^-$ ion is C_1 and $KXeO_3F$ belongs to space group C_{2v}^9, its vibrational spectra are simple and can be analyzed in terms of

C_{3v} symmetry of the isolated anion [167]. Accordingly, there should be six fundamental modes (3 a_1 + 3 e), all infrared and Raman active. Assignments for the $[XeO_3F]^-$ ion are shown in Table 26. There is considerable mixing between the

Table 26. Observed Raman Frequencies (cm^{-1}) and Assignments for $[XeO_3F]^-$ [167]

$[XeO_3F]^-$	Assignment Description
781	$\nu_1(a_1)$, ν_s(Xe—O)
358	$\nu_2(a_1)$, δ_s(O—Xe—O)
257	$\nu_3(a_1)$, ν_s(Xe—F) (bridge)
822	ν_4(e), ν_{as}(Xe—O)
315	ν_5(e), δ(O—Xe—O) degen.
222	ν_6(e), δ(O—Xe—F) degen.

$\nu_2(a_1)$ and $\nu_3(a_1)$ modes and a normal coordinate analysis was undertaken to differentiate between them. The $\nu_3(a_1)$ mode is really an Xe—F—Xe bridging vibration and thus considerably lower in frequency than the usual Xe—F stretching vibrations. It was, therefore, not observed in the original work [145]. The corresponding frequency ν_3(Xe—Cl) in the $[XeO_3Cl]^-$ salts occurs at 250 cm^{-1}, pointing to much weaker Xe—Cl—Xe bridging, since this vibration would be expected to be lowered to 217 cm^{-1}. This is borne out by the observation of long (2.95 Å) Xe—Cl bridging bond lengths [166].

IV $[XeOF_3]^-$ and $[XeO_2F_3]^-$

Anions derived from the unstable xenon oxide fluorides, $XeOF_2$ and XeO_2F_2, have been reported [168]. $CsXeOF_3$ is prepared by pumping the solvent from an HF solution containing solid $XeOF_2$ and CsF, while allowing it to warm slowly from −196 to 0 °C. The bright yellow salt, $Cs^+[XeOF_3]^-$, is stable at room temperature for several hours. If the solution is allowed first to warm slowly to room temperature followed by removal of HF under reduced pressure $Cs^+[XeO_2F_3]^-$ is obtained in quantitative yield instead. This results from the disproportionation of $XeOF_2$ [169] according to the equations (Eqs. 45, 46) [188].

$$2\,XeOF_2 \rightarrow XeO_2F_2 + XeF_2 \tag{45}$$

$$XeO_2F_2 + F^- \xrightarrow{HF} [XeO_2F_3]^- \tag{46}$$

The Raman spectra for $[XeOF_3]^-$ (XIX) and $[XeO_2F_3]^-$ (XX) are given in Table 27.

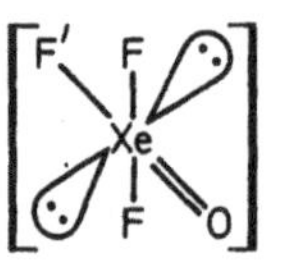

(XIX)

(XX)

Those of $[XeOF_3]^-$ were assigned on the basis of C_{2v} symmetry but only those of the F—Xe=O bending mode are reasonably reliable [168].

Table 27. Observed Raman Frequencies (cm^{-1}) and Assignments for $[XeOF_3]^-$ and $[XeO_2F_3]^-$ [168]

$[XeOF_3]^-$	$[XeO_2F_3]^-$	Assignment Description
	861 (16) 857 (8)	b_2, $Xe{=}^{16}O$ asym str.
768 (7)	834 (100)	a_1, $Xe{=}^{16}O$ sym str.
727 (32)		a_1, $Xe{=}^{18}O$ sym str.
503 (61)	541 (2)	a_1, Xe—F′ str.
487 (10)	514 (18)	b_1, XeF_2 asym str.
464 (100)	459 (25)	a_1, XeF_2 sym str.
381 (5)		b_2, $F'{-}Xe{=}^{16}O$ out-of-plane bend
370 (6)		b_2, $F{-}Xe{=}^{18}O$ out-of-plane bend
292 (1)		a_1, sym $F'F_2XeO$ in-plane bend
270 (5)		b_1, $Xe{=}^{16}O$ wag
260 (8)		b_1, $Xe{=}^{18}O$ wag
219 (1)		b_2, XeF_2 out-of-plane bend
167 (6)		b_1, Xe—F wag or XeF_2 in-plane bend

Note added in proof:

The salt NF_4XeF_7 containing the anion XeF_7^- has been synthesized [172]. It decomposes according to

$$NF_4XeF_7 \rightarrow NF_3 + F_2 + XeF_6$$

There is evidence for the existence of $(NF_4)_2XeF_8$ as an intermediate, but this has not been isolated.

The enthalpy of formation of NF_4XeF_7 was calculated to be —491 kJ/mole^{-1}, and it is thus the most energetic NF_4^+ salt known.

D References

1. Holloway, J. H.: Noble-Gas Chemistry, London, Methuen, 1968.
   Hoppe, R.: Die Chemie der Edelgase, Top. Curr. Chem. *5*, 213–346 (1965)
2. Bartlett, N., Sladky, F. O.: Noble-Gas Chemistry, in: Comprehensive Inorganic Chemistry (ex. ed. Trotman-Dickenson, A. F.) p. 229, Oxford, Pergamon, 1973
3. Bartlett, N., Sladky, F. O.: Noble-Gas Chemistry, in: Comprehensive Inorganic Chemistry (ex. ed. Trotman-Dickenson, A. F.) p. 213, Oxford, Pergamon, 1973
4. Gillespie, R. J., Landa, B., Schrobilgen, G. J.: J. Inorg. Nucl. Chem., Suppl., 179 (1976)
5. Cohen, B., Peacock, R. D.: J. Inorg. Nucl. Chem. *28*, 3056 (1966)
6. Stein, L., et al.: J. Chem. Soc., Chem. Commun., 502 (1978)
7. Stein, L., Henderson, W. W.: J. Am. Chem. Soc. *102*, 2856 (1980)
8. Stein, L.: J. Fluorine Chem. *20*, 65 (1982)
9. Howard, W. F., Andrews, L.: J. Am. Chem. Soc. *97*, 2956 (1975)
10. Brundle, C. R., Jones, G. R.: J. Chem. Soc., Faraday Trans. 2, *68*, 959 (1972)
11. Gunn, S. R.: J. Phys. Chem. *71*, 2934 (1967)
12. Berkowitz, J., Holloway, J. H.: J. Chem. Soc., Faraday Trans. 2, *74*, 2077 (1978)
13. Johnston, H. S., Woolfolk, R.: J. Chem. Phys. *41*, 269 (1964)
14. Berkowitz, J., et al.: ibid. *75*, 1461 (1971)
15. Berkowitz, J., Chupka, W. A.: Chem. Phys. Lett. *7*, 447 (1970)
16. Selig, H., Peacock, R. D.: J. Am. Chem. Soc. *86*, 3895 (1964)

17. Edwards, A. J., Holloway, J. H., Peacock, R. D.: Proc. Chem. Soc. 275 (1963)
18. Frlec, B., Holloway, J. H.: J. Chem. Soc., Chem. Commun., 370 (1973)
19. Frlec, B., Holloway, J. H.: ibid. 89 (1974)
20. Gillespie, R. J., Schrobilgen, G. J.: ibid. 90 (1974)
21. Holloway, J. H., Schrobilgen, G. J.: ibid. 623 (1975)
22. Gillespie, R. J., Schrobilgen, G. J.: Inorg. Chem. *15*, 22 (1976)
23. Frlec, B., Holloway, J. H.: ibid. *15*, 1263 (1976)
24. Holloway, J. H., et al.: J. Chem. Phys. *66*, 2627 (1977)
25. Sokolov, V. B., Tsinoev, V. G., Ryshkov, A. V.: Teor. Eksp. Khim. *16*, 245 (1980)
26. Gillespie, R. J., Martin, D., Schrobilgen, G. J.: J. Chem. Soc., Dalton Trans., 1898 (1980)
27. Maslov, O. D., et al.: Russ. J. Phys. Chem. (Engl. Transl.) *41*, 984 (1967)
28. Binenboym, J., Selig, H., Shamir, J.: J. Inorg. Nucl. Chem. *30*, 2863 (1968)
29. Holloway, J. H., Knowles, J. G.: J. Chem. Soc. (A), 756 (1969)
30. McRae, V. M., Peacock, R. D., Russell, D. R.: J. Chem. Soc., Chem. Commun., 62 (1969)
31. Burgess, J., et al.: J. Inorg. Nucl. Chem., Suppl., 183 (1976)
32. Sladky, F. O., Bulliner, P. A., Bartlett, N.: J. Chem. Soc. (A), 2179 (1969)
33. Gillespie, R. J., Landa, B.: Inorg. Chem. *12*, 1383 (1973)
34. Bartlett, N., et al.: ibid. *12*, 1717 (1973)
35. Fuggle, J. C., et al.: J. Chem. Soc., Dalton Trans., 205 (1974)
36. Burgess, J., Frlec, B., Holloway, J. H.: ibid. 1740 (1974)
37. Bartlett, N., et al.: Inorg. Chem. *13*, 780 (1974)
38. Frlec, B., Holloway, J. H.: J. Chem. Soc., Dalton Trans., 535 (1975)
39. Zalkin, A., et al.: Inorg. Chem. *17*, 1318 (1978)
40. de Waard, H., et al.: J. Chem. Phys. *70*, 3247 (1979)
41. Frlec, B., Holloway, J. H.: J. Inorg. Nucl. Chem., Suppl., 167 (1976)
42. Fawcett, J., Frlec, B., Holloway, J. H.: J. Fluorine Chem. *8*, 505 (1976)
43. Gillespie, R. J., Netzer, A., Schrobilgen, G. J.: Inorg. Chem. *13*, 1455 (1974)
44. Holloway, J. H., et al.: C. R. Acad. Sci. Paris, Ser. C., *282*, 519 (1976)
45. Schrobilgen, G. J., et al.: Inorg. Chem. *17*, 980 (1978)
46. Holloway, J. H., Schrobilgen, G. J., Taylor, P.: J. Chem. Soc., Chem. Commun., 40 (1975)
47. Tucker, P. A., et al.: Acta. Crystallogr., Sect. B., *31*, 906 (1975)
48. Holloway, J. H., Schrobilgen, G. J.: Inorg. Chem. *19*, 2632 (1980)
49. Holloway, J. H., Schrobilgen, G. J.: ibid. *20*, 3363 (1981)
50. Kaul, W., Ruchs, R.: Z. Naturforsch., Teil A, *15*, 326 (1960)
51. Henglein, A., Muccini, G. A.: Angew. Chem. *72*, 630 (1960)
52. Liebman, J. F., Allen, L. C.: J. Am. Chem. Soc. *92*, 3539 (1970)
53. Liebman, J. F., Allen, L. C.: J. Chem. Soc., Chem. Commun., 1355 (1969)
54. Allen, L. C., Huheey, J. E.: J. Inorg. Nucl. Chem. *42*, 1523 (1980)
55. Schreiner, F., Malm, J. G., Hindman, J. C.: J. Am. Chem. Soc. *87*, 25 (1965)
56. Prusakov, V. N., Sokolov, V. B.: At. Energ. *31*, 259 (1971)
57. Prusakov, V. N., Sokolov, V. B.: Zh. Fiz. Khim. *45*, 2950 (1971)
58. McKee, D. E., et al.: J. Chem. Soc., Chem. Commun., 26 (1973)
59. Gillespie, R. J., Schrobilgen, G. J.: Inorg. Chem. *13*, 1230 (1974)
60. Klimov, V. D., Legasov, V. A., Khoroshev, S. S.: Zh. Fiz. Khim. *52*, 1790 (1978)
61. Žemva, B., Slivnik, J., Šmalc, A.: J. Fluorine Chem. *6*, 191 (1975)
62. Lui, B., Schaefer, H. F.: J. Chem. Phys. *55*, 2369 (1971)
63. Sladky, F. O., et al.: J. Chem. Soc., Chem. Commun., 1048 (1968)
64. Gillespie, R. J., Schrobilgen, G. J.: Inorg. Chem. *13*, 765 (1974)
65. Gillespie, R. J., Landa, B., Schrobilgen, G. J.: J. Chem. Soc., Chem. Commun., 1543 (1971)
66. Gillespie, R. J., Landa, B., Schrobilgen, G. J.: ibid. 607 (1972)
67. Lentz, D., Seppelt, K.: Angew. Chem., Int. Ed. Engl. *18*, 66 (1979)
68. Bartlett, N., Jha, N. K.: The Xenon-Platinum Hexafluoride Reaction and Related Reactions in Noble Gas Compounds (ed. Hyman, H. H.) p. 23, Chicago, University of Chicago Press, 1963
69. Edwards, A. J., Holloway, J. H., Peacock, R. D.: Some Properties of Xenon Fluorides, in Noble Gas Compounds (ed. Hyman, H. H.) p. 71, Chicago, University of Chicago Press, 1963

70. Sladky, F. O., et al.: J. Chem. Soc., Chem. Commun., 1048 (1968)
71. Legasov, V. A., Prusakov, V. N., Chaivanov, B. B.: Russ. J. Phys. Chem. (Engl. Transl.) *44*, 1496 (1970)
72. Legasov, V. A., Chaivanov, B. B.: ibid. *45*, 325 (1971)
73. Chaivanov, B. B., et al.: I.A.E. Report 2186, Moscow 1972
74. Stein, L.: Nature (London) *243*, 30 (1973)
75. Žemva, B., Slivnik, J.: J. Inorg. Nucl. Chem., Suppl., 173 (1976)
76. Bartlett, N., Sladky, F. O.: J. Am. Chem. Soc. *90*, 5316 (1968)
77. Christe, K. O., Wilson, R. D.: Inorg. Nucl. Chem. Lett. *9*, 845 (1973)
78. Bartlett, N., Sladky, F. O.: Noble Gas Chemistry, in Comprehensive Inorganic Chemistry (ex. ed. Trotman-Dickenson, A. F.) p. 277, Oxford, Pergamon, 1973
79. Meinert, H., Rudiger, S.: Z. Chem. *9*, 35 (1969)
80. Meinert, H., Rudiger, S.: ibid. *9*, 71 (1969)
81. Jones, G. R., Burbank, R. D., Bartlett, N.: Inorg. Chem. *9*, 2264 (1970)
82. Burns, J. H., Ellison, R. D., Levy, H. A.: Acta Cryst. *18*, 11 (1965)
83. Bartlett, N., Wechsberg, M.: Z. Anorg. Allg. Chem. *385*, 5 (1971)
84. Klimov, V. D., Marinin, A. S.: Zh. Prikl. Spektrosk *21*, 184 (1974)
85. Legasov, V. N., Prusakhov, V. N., Chaivanov, B. B.: I.A.E. Report 2185, Moscow 1972
86. Zarubin, V. N., Marinin, A. S.: Russ. J. Inorg. Chem. (Engl. Transl.) *19*, 1599
87. Žemva, B., Slivnik, J., Bohinc, M.: J. Inorg. Nucl. Chem. *38*, 73 (1976)
88. Bartlett, N., et al.: J. Chem. Soc., Chem. Commun., 703 (1969)
89. Eisenberg, M., DesMarteau, D. D.: Inorg. Nucl. Chem. Lett. *6*, 29 (1970)
90. Eisenberg, M., DesMarteau, D. D.: Inorg. Chem. *11*, 1901 (1972)
91. Bartlett, N., et al.: ibid. *11*, 1124 (1972)
92. Seppelt, K.: Angew. Chem., Int. Ed. Engl. *11*, 723 (1972)
93. Seppelt, K., Rupp, H. H.: Z. Anorg. Allg. Chem. *409*, 338 (1974)
94. Sladky, F. O.: Angew. Chem., Int. Ed. Engl. *8*, 373 (1969)
95. Sladky, F. O.: Monatsh. *101*, 1571 (1970)
96. LeBlond, R. D., DesMarteau, D. D.: J. Chem. Soc., Chem. Commun., 555 (1974)
97. Sawyer, J. F., Schrobilgen, G. J., Sutherland, S. J.: Inorg. Chem. *21*, 4064 (1982)
98. Bartlett, N., Lohmann, D. H.: Proc. Chem. Soc., 115 (1962)
99. Bartlett, N.: ibid. 218 (1962)
100. Chernick, C. L., et al.: Science *138*, 136 (1962)
101. Fields, P. R., Stein, L., Zirin, M. H.: J. Am. Chem. Soc. *84*, 4164 (1962)
102. Stein, L.: U.S.P. 3 660 300, 1972
103. Stein, L.: J. Am. Chem. Soc. *91*, 5396 (1969)
104. Stein, L.: Science *168*, 362 (1970)
105. Stein, L.: Yale Scientific Mag. *44* (8), 2 (1970)
106. Pitzer, K. S.: J. Chem. Soc., Chem. Commun., 760 (1975)
107. Stein, L.: Science *175*, 1463 (1972)
108. Stein, L.: J. Inorg. Nucl. Chem. *35*, 39 (1973)
109. McKee, D. E., Adams, C. J., Bartlett, N.: Inorg. Chem. *12*, 1722 (1973)
110. Gillespie, R. J., Schrobilgen, G. J.: ibid. *13*, 2370 (1974)
111. Gillespie, R. J., Landa, B., Schrobilgen: ibid. *15*, 1256 (1976)
112. Boldrini, P., et al.: ibid. *13*, 1690 (1974)
113. McKee, D. E., Zalkin, A., Bartlett, N.: ibid. *12*, 1713 (1973)
114. Gillespie, R. J., et al.: J. Chem. Soc., Dalton Trans., 2234 (1977)
115. Gillespie, R. J.: Molecular Geometry, London, Van Nostrand-Reinhold, 1972
116. Selig, H.: Science *144*, 537 (1964)
117. Adams, C. J., Bartlett, N.: Isr. J. Chem. *17*, 114 (1978)
118. Hyman, H. H., Quarterman, L. A., Hydrogen Fluoride Solutions Containing Xenon Difluoride, Xenon Tetrafluoride and Xenon Hexafluoride, in: Noble Gas Compounds (ed. Hyman, H. H.) p. 275, Chicago, University of Chicago Press, 1963
119. Rothman, M. J., et al.: J. Chem. Phys. *73*, 375 (1980)
120. Leary, K., Zalkin, A., Bartlett, N.: Inorg. Chem. *13*, 775 (1974)
121. Žemva, B., Miličev, S., Slivnik, J.: J. Fluorine Chem. *11*, 519 (1978)
122. Gard, G. L., Cady, G. H.: Inorg. Chem., *3*, 1745 (1964)

123. DesMarteau, D. D., Eisenberg, M.: ibid. *11*, 2641 (1972)
124. Hanžel, D.: Inorg. Nucl. Chem. Lett. *12*, 539 (1976)
125. Christe, K. O., Curtis, C. C., Wilson, R. D.: J. Inorg. Nucl. Chem., Suppl., 159 (1976)
126. Žemva, B., Zupan, J., Slivnik, J.: J. Inorg. Nucl. Chem. *35*, 3941 (1973)
127. Moody, G. J., Selig, H.: ibid. *28*, 2429 (1966)
128. Žemva, B., Slivnik, J.: J. Fluorine Chem. *8*, 369 (1976)
129. Aubert, J., Cady, G. H.: Inorg. Chem. *9*, 2600 (1970)
130. Bartlett, N., et al.: J. Chem. Soc. (A), 1190 (1967)
131. Slivnik, J., et al.: J. Inorg. Nucl. Chem. *32*, 1397 (1970)
132. Frlec, B., et al.: ibid. *34*, 2938 (1972)
133. Bohinc, M., Frlec, B.: ibid. *34*, 2942 (1972)
134. Pullen, K. E., Cady, G. H.: Inorg. Chem. *6*, 2267 (1967)
135. Leary, K., et al.: ibid. *12*, 1726 (1973)
136. Pullen, K. E., Cady, G. H.: ibid. *6*, 1300 (1967)
137. Pullen, K. E., Cady, G. H.: ibid. *5*, 2057 (1966)
138. Žemva, B., Miličev, S., Slivnik, J.: J. Fluorine Chem. *11*, 545 (1978)
139. Rundle, R. E.: J. Am. Chem. Soc. *85*, 112 (1963)
140. Pimentel, G. C., Spratley, R. D.: ibid. *85*, 826 (1963)
141. Burbank, R. D., Jones, G. R.: Science *168*, 248 (1970)
142. Burbank, R. D., Jones, G. R.: J. Am. Chem. Soc. *96*, 43 (1974)
143. Frame, H. D.: Chem. Phys. Lett. *3*, 182 (1969)
144. Leary, K., Bartlett, N.: J. Chem. Soc., Chem. Commun., 903 (1972)
145. Selig, H.: Inorg. Chem. *5*, 183 (1966)
146. Waldman, M. C., Selig, H.: J. Inorg. Nucl. Chem. *35*, 2173 (1973)
147. Sladky, F. O.: Monatsh. *101*, 1578 (1970)
148. Sladky, F. O.: Angew. Chem., Int. Ed. Engl. *9*, 375 (1970)
149. Keller, N., Schrobilgen, G. J.: Inorg. Chem. *20*, 2118 (1981)
150. Wechsberg, M., et al.: ibid. *11*, 3063 (1972)
151. Gillespie, R. J., Schrobilgen, G. J.: ibid. *13*, 1694 (1974)
152. Gillespie, R. J., Schrobilgen, G. J., Slim, D. R.: J. Chem. Soc., Dalton Trans., 1003 (1977)
153. DesMarteau, D. D.: J. Am. Chem. Soc. *100*, 6270 (1978)
154. Peacock, R. D., Selig, H., Sheft, I.: Proc. Chem. Soc., 285 (1964)
155. Peacock, R. D., Selig, H., Sheft, I.: J. Inorg. Nucl. Chem. *28*, 2561 (1966)
156. Moody, G. J., Selig, H.: Inorg. Nucl. Chem. Lett. *2*, 319 (1966)
157. Peterson, S. W., et al.: Science *173*, 1238 (1971)
158. Wang, S. Y., Lohr, L. L.: J. Chem. Phys. *60*, 3916 (1974)
159. Holloway, J. H., Selig, H., El-Gad, U.: J. Inorg. Nucl. Chem. *35*, 3624 (1973)
160. Schrobilgen, G. J., et al.: J. Chem. Soc., Chem. Commun., 894 (1980)
161. Christe, K. O.: Inorg. Chem. *11*, 1215 (1972)
162. Kaučič, V., et al.: submitted for publication
163. Hodgson, D. J., Ibers, J. A.: Inorg. Chem. *8*, 326 (1969)
164. Jaselskis, B., Huston, J. L., Spittler, T. M.: J. Am. Chem. Soc. *91*, 1874 (1969)
165. Jaselskis, B., Spittler, T. M., Huston, J. L.: ibid. *89*, 2790 (1967)
166. Willet, R. D., Peterson, S. W., Coyle, B. A.: ibid. *99*, 8202 (1977)
167. La Bonville, P., Ferraro, J. R., Spittler, T. M.: J. Chem. Phys. *55*, 631 (1975)
168. Gillespie, R. J., Schrobilgen, G. J.: J. Chem. Soc., Chem. Commun., 595 (1977)
169. Claassen, H. H., et al.: J. Chem. Phys. *49*, 253 (1969)
170. Bartlett, N., et al.: J. Chem. Soc., Chem. Comm., 550 (1966)
171. Žemva, B., Slivnik, J.: J. Inorg. Nucl. Chem. *33*, 3953 (1971)
172. Christe, K. O., et al.: J. Fluor. Chem. *23*, 399 (1983)

Extraction of Metals from Sea Water

Klaus Schwochau

Institute of Chemistry, Nuclear Research Centre (KFA) D-5170 Jülich, FRG

Table of Contents

1 Introduction

The oceans represent an almost inexhaustible source of raw materials. About 80 elements could be detected in sea water till now; their molar concentrations differ by more than twenty orders of magnitude. However, the chemistry of sea water is dominated only by the presence of the six ions Cl^-, Na^+, Mg^{2+}, Ca^{2+}, K^+, and SO_4^{2-}, which constitute more than 99.5% of the dissolved material [1]. Open ocean water contains a total of dissolved salts within the range of 33 to 37 g/l [2].

The metals of the elements occuring in the oceans are the concern of this article. Metal ions dissolved in sea water are compiled in Table 1.

Table 1. Concentration and chemical species of metal ions dissolved in sea water [1,3]

Metal	Concentration		Main chemical species
	[mol/l]	[mg/l]	
Li	$2.5 \cdot 10^{-5}$	0.173	Li^+
Be	$6.3 \cdot 10^{-10}$	$5.6 \cdot 10^{-6}$	$Be(OH)^+$
Na	0.468	$10.77 \cdot 10^3$	Na^+
Mg	$5.32 \cdot 10^{-2}$	$12.9 \cdot 10^2$	Mg^{2+}
Al	$3.7 \cdot 10^{-8}$	$1.0 \cdot 10^{-3}$	$[Al(OH)_4]^-$
K	$1.02 \cdot 10^{-2}$	$3.8 \cdot 10^2$	K^+
Ca	$1.02 \cdot 10^{-2}$	$4.12 \cdot 10^2$	Ca^{2+}
Sc	$1.3 \cdot 10^{-11}$	$6 \cdot 10^{-7}$	$Sc(OH)_3$
Ti	$2 \cdot 10^{-8}$	$1 \cdot 10^{-3}$	$Ti(OH)_4$
V	$3.7 \cdot 10^{-8}$	$1.9 \cdot 10^{-3}$	$H_2VO_4^-$, HVO_4^{2-}
Cr	$1.54 \cdot 10^{-9}$	$8.0 \cdot 10^{-5}$	$Cr(OH)_3$, CrO_4^{2-}
Mn	$3.6 \cdot 10^{-9}$	$2.0 \cdot 10^{-4}$	Mn^{2+}, $MnCl^+$
Fe	$2.3 \cdot 10^{-8}$	$1.3 \cdot 10^{-3}$	$[Fe(OH)_2]^+$, $[Fe(OH)_4]^-$
Co	$6.8 \cdot 10^{-10}$	$4.0 \cdot 10^{-5}$	Co^{2+}
Ni	$3.4 \cdot 10^{-9}$	$2.0 \cdot 10^{-4}$	Ni^{2+}
Cu	$1.6 \cdot 10^{-9}$	$1.0 \cdot 10^{-4}$	$CuCO_3$, $Cu(OH)^+$
Zn	$1.5 \cdot 10^{-10}$	$1.0 \cdot 10^{-5}$	$Zn(OH)^+$, Zn^{2+}, $ZnCO_3$
Ga	$4.3 \cdot 10^{-10}$	$3 \cdot 10^{-5}$	$[Ga(OH)_4]^-$
Ge	$6.9 \cdot 10^{-10}$	$5 \cdot 10^{-5}$	$Ge(OH)_4$
Rb	$1.4 \cdot 10^{-6}$	0.120	Rb^+
Sr	$9.2 \cdot 10^{-5}$	8.1	Sr^{2+}
Y	$1.5 \cdot 10^{-11}$	$1.3 \cdot 10^{-6}$	$Y(OH)_3$
Zr	$3.3 \cdot 10^{-10}$	$3 \cdot 10^{-5}$	$Zr(OH)_4$
Nb	$1 \cdot 10^{-10}$	$1 \cdot 10^{-5}$	—
Mo	$1.0 \cdot 10^{-7}$	$1.0 \cdot 10^{-2}$	MoO_4^{2-}
Ag	$9.3 \cdot 10^{-11}$	$1.0 \cdot 10^{-5}$	$[AgCl_2]^-$
Cd	$8.9 \cdot 10^{-11}$	$1.0 \cdot 10^{-5}$	$CdCl_2$
In	$0.8 \cdot 10^{-12}$	$1 \cdot 10^{-7}$	$In(OH)_2^+$
Sn	$8.4 \cdot 10^{-11}$	$1 \cdot 10^{-5}$	$[SnO(OH)_3]^-$
Sb	$1.7 \cdot 10^{-9}$	$2.1 \cdot 10^{-4}$	$[Sb(OH)_6]^-$
Cs	$3 \cdot 10^{-9}$	$4.0 \cdot 10^{-4}$	Cs^+
Ba	$1.5 \cdot 10^{-7}$	$2.0 \cdot 10^{-2}$	Ba^{2+}
La	$2 \cdot 10^{-11}$	$3 \cdot 10^{-6}$	$La(OH)_3$
Ce	$1 \cdot 10^{-10}$	$1 \cdot 10^{-6}$	$Ce(OH)_3$
Pr	$4 \cdot 10^{-12}$	$6 \cdot 10^{-7}$	$Pr(OH)_3$
Nd	$1.9 \cdot 10^{-11}$	$3 \cdot 10^{-6}$	$Nd(OH)_3$
Sm	$3 \cdot 10^{-12}$	$5 \cdot 10^{-8}$	$Sm(OH)_3$

Table 1. (continued)

Metal	Concentration		Main chemical species
	[mol/l]	[mg/l]	
Eu	$9 \cdot 10^{-13}$	$1 \cdot 10^{-8}$	$Eu(OH)_3$
Gd	$4 \cdot 10^{-12}$	$7 \cdot 10^{-7}$	$Gd(OH)_3$
Tb	$9 \cdot 10^{-13}$	$1 \cdot 10^{-7}$	$Tb(OH)_3$
Dy	$6 \cdot 10^{-12}$	$9 \cdot 10^{-7}$	$Dy(OH)_3$
Ho	$1 \cdot 10^{-12}$	$2 \cdot 10^{-7}$	$Ho(OH)_3$
Er	$4 \cdot 10^{-12}$	$8 \cdot 10^{-7}$	$Er(OH)_3$
Tm	$8 \cdot 10^{-13}$	$2 \cdot 10^{-7}$	$Tm(OH)_3$
Yb	$5 \cdot 10^{-12}$	$8 \cdot 10^{-7}$	$Yb(OH)_3$
Lu	$9 \cdot 10^{-13}$	$2 \cdot 10^{-7}$	$Lu(OH)_3$
Hf	$4 \cdot 10^{-11}$	$7 \cdot 10^{-6}$	—
Ta	$1 \cdot 10^{-11}$	$2 \cdot 10^{-6}$	—
W	$5 \cdot 10^{-10}$	$1.0 \cdot 10^{-4}$	WO_4^{2-}
Re	$2 \cdot 10^{-11}$	$4.0 \cdot 10^{-6}$	ReO_4^-
Au	$5.1 \cdot 10^{-11}$	$1 \cdot 10^{-5}$	$[AuCl_2]^-$
Hg	$5.0 \cdot 10^{-11}$	$1 \cdot 10^{-5}$	$[HgCl_4]^{2-}$, $HgCl_2$
Tl	$5 \cdot 10^{-11}$	$1 \cdot 10^{-5}$	—
Pb	$7.2 \cdot 10^{-11}$	$1.5 \cdot 10^{-5}$	$PbCO_3$, $[Pb(CO_3)_2]^{2-}$
Bi	$1 \cdot 10^{-10}$	$2 \cdot 10^{-5}$	BiO^+, $Bi(OH)_2^+$
Ra	$3 \cdot 10^{-16}$	$7 \cdot 10^{-11}$	Ra^{2+}
Th	$4 \cdot 10^{-11}$	$1 \cdot 10^{-5}$	$Th(OH)_4$
Pa	$2 \cdot 10^{-16}$	$5 \cdot 10^{-11}$	—
U	$1.4 \cdot 10^{-8}$	$3.3 \cdot 10^{-3}$	$[UO_2(CO_3)_3]^{4-}$

The five metal ions Na^+, Mg^{2+}, Ca^{2+}, K^+, and Sr^{2+} which occur at concentrations down to 1 mg/l belong to the major constituents, since they contribute significantly to the salinity. Sodium, magnesium, and potassium are found in sea water in nearly constant proportions [2] whereas calcium shows statistically a higher concentration in deeper waters than in surface waters by about 0.5%. In the case of strontium the mean Sr:Cl ratio seems to be lower in the surface layers by up to 3% [4].

Most trace metals in sea water occur in the concentration range of 0.01 to 10 µg/l; lower concentrations of metals as far as mentioned in Table 1, are found for Be, Sc, Y, In, the lanthanoids, Hf, Ta, Re, and finally for the extremely low concentrated actinoids Ra and Pa. The concentration of any trace metal in sea water normally does not vary by as much as one order of magnitude, but very precise analyses are required to determine their distribution and chemical speciation [1].

Only a few metals belonging to the major constituents can nowadays be economically extracted from sea water; sodium, magnesium, calcium, and to a minor extent potassium are recovered in industrial quantities. Nevertheless, procedures have been developed or proposed for the extraction of many of the trace metals [5]. It has been estimated that in view of capital and energy requirements for pumping sea water to a collection site or circulating it through process equipment the recovery of any metal lower in concentration than strontium (8.1 mg/l) cannot be profitable [6]. Unquestionably, however, the actual value of a metal to be extracted from sea water has to be taken into account.

Several comprehensive reviews on the extraction of inorganic materials from sea water have already been published. Tallmadge, Butt, and Solomon [7)] evaluated methods for recovery of different minerals. McIlhenny has reviewed the extraction of raw materials and economic inorganic materials from sea water [5)]. The recovery of minerals from the oceans is also discussed by Hanson and Murthy [9)], Seetharam and Srinivasan [10)], and very recently by Massie [11)] and Ogata [12)].

The basic methods are being used or have been proposed for separation of mineral products from sea water, comprise sorption, ion exchange, solvent extraction, flotation, precipitation, evaporation, distillation, electrolysis, and electrodialysis [7,13)], In this article emphasis is given to such metal extraction methods which are based on complex formation between suitable functional groups of the host and metal ions as guest.[1] Mere physical methods like evaporation, distillation, and electrolysis as primary separation steps are only mentioned incidentally. Moreover, extractions of metals mainly from natural sea water, brines, and bitterns will be reviewed, whereas experiments on metal extractions from artificial sea water were of less interest in this context. The metals discussed are divided into alkali, alkaline earth, trace metals, and uranium and are treated in order to their atomic number. Although uranium formally belongs to the trace metals the extensive investigations on the extraction of uranium from sea water with regard to nuclear energy supply made it necessary to devote an entire chapter to this element.

2 Metal Extraction

2.1 Alkali Metals

2.1.1 Lithium

With regard to the anticipated power supply by controlled thermonuclear reactors in the next century, the demand for lithium as blanket material for breeding tritium and for cooling the reactor is expected to grow. The demand varies depending on the reactor type; the quantity of natural lithium required for a solid lithium blanket, is about 5.5–46.8 kg/MWe, for a liquid lithium blanket the quantity required may be as high as about 1000 kg/MWe [14)]. Commercially lithium is mainly used for ceramics, glass, multipurpose greases, metallurgy, air conditioning, catalysers, and pharmaceuticals. Moreover, it has been proposed as a material for high performance batteries powering electric vehicles and storing off-peak energy in electric utility systems [15)]. The most important conventional lithium sources are the minerals amblygonite (Li, Na) Al(F, OH) PO_4, spodumene $LiAlSi_2O_6$, petalite $LiAlSi_4O_{10}$, and lepidolite K(Li, Al)$_3$ [Al, Si]$_4O_{10}$(F, OH)$_2$. However, in order to meet the growing energy-related lithium demand in future, brines have received more attention and sea water is examined as an alternate huge source for lithium.

The lithium concentration in sea water is almost constant; according to Table 1 the concentration amounts to 0.173 mg/l, however, the total quantity occuring in sea water is tremendous and estimated to be 2.4×10^{11} tons.

---

1 Solvent extraction based on crown compounds, see the contribution by Takeda, Y., this series 121.

The following process concept, a combination of solar evaporation and ion exchange, for lithium extraction from sea water, has been proposed by Steinberg and Dang [14]. Sea water flows into the solar pond by tidal waves. During the process of fractional crystallization, sodium chloride, calcium sulfate, and magnesium chloride will precipitate first. When the sea water is concentrated to 10^{-6} parts of the original volume by evaporation, lithium chloride will begin to precipitate. The concentrated lithium brine is passed through an ion exchange bed of Dowex 50-X16 consisting of beads of polystyrene crosslinked with 16% divinylbenzene. The Dowex resin is selective for exchange of its hydrogen ions with the cations of the sea water in the order of K^+, Na^+, Li^+, and Mg^{2+}. Lithium ions are first eluted from the pregnant bed with 0.2–0.5 M hydrochloric acid. The eluted LiCl solution flows into the evaporator to be concentrated or is recycled back to the bed in order to elute more lithium ions. LiCl is then transferred into an electrolyzer where lithium metal is produced.

Other processes for separation of lithium from sea water have been developed. By using the ion-retardation resin, Retardion Ag11 A8, a divinylbenzene cross-linked copolymer of polyacrylic acid and polystyrene containing the weakly acidic $—CH_2COOH$ and strongly basic $—CH_2N^+(CH_3)_3Cl^-$ as functional groups, lithium ions were brought to a 30 fold concentration compared to the level found in sea water and could be eluted with dilute ethanol [16].

Amorphous aluminum oxide has recently been proved to extract lithium from brines and bitterns having lithium concentrations of 0.83 and 13.1 mg/l, respectively. The sorption may be explained by the formation of hydrous lithium aluminum oxide. The sorption capacity of amorphous hydrous aluminum oxide was found to be 4.0 mmol/g. For brines and bitterns the lithium concentration factors on the sorbent attained values of 370 and 130, respectively; equilibrium was reached after 7 days. The desorption of lithium ions was carried out with boiling water yielding a maximum concentration factor of lithium in the eluate of 46 in reference to the initial lithium concentration of the brines. Lithium was separated from the eluates by solvent extraction with cyclohexane containing thenoyltrifluoracetone and trioctylphosphine oxide, subsequent back extraction with hydrochloric acid, and precipitation of lithium phosphate by addition of K_3PO_4. The purity of the precipitate amounted to at least 95% [17–21].

Instead of hydrous aluminum oxide, aluminum foils have very recently been proposed for the extraction of lithium from sea water. However, not the aluminium metal itself, but its corrosion product hydrous aluminum oxide formed in sea water on the metal surface was found to be the effective lithium binding agent [15]; thus, there seems to be no fundamental difference compared to the direct application of pure hydrous aluminum oxide.

2.1.2 Sodium

Sodium is the most abundant metal in sea water. Sodium chloride is commercially produced from sea water by solar evaporation. Salt is a dietary necessity, but only a small fraction of the production is actually used as 'table salt' in foods. The chemical usages for sodium are so extensive that salt is one of the most important raw materials for the chemical industries.

Aside from the solar evaporation method, precipitation [22-25], electrolytic [26-31], and ion exchange [32,33] processes have been tested for sodium recovery, but only on a small scale; in contrast, electrodialysis has been more significant.

The procedure of electrodialysis [7] proposed for recovery of Na_2CO_3, uses sea water purified from magnesium, which is fed to the anode while sodium bicarbonate is supplied to the cathode. The two electrodes are separated by a sulfonated anion resin. The net transfer of sodium to the cathode produces a saturated solution of Na_2CO_3. Bicarbonate is regenerated and recycled from a portion of the carbonate product [28,29].

Substantial quantities of sodium chloride have been produced in Japan from sea water concentrates using other types of electrodialysis equipment; the total output of such electrodialytic salt plants has exceeded 260000 tons per year [5]. An electrodialysis stack consists of multiple pairs of ion exchange membranes, one membrane being permeable to anions (A) the other to cations (C) (Fig. 1).

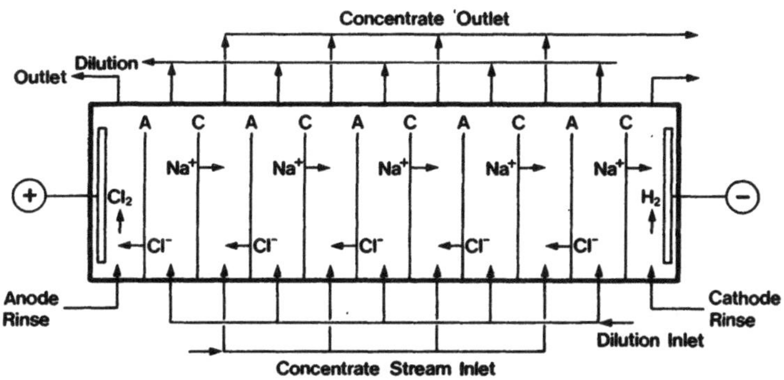

Fig. 1. Electrodialysis salt concentration from sea water [5]

When an electrical potential gradient is imposed on the stack, alternate compartments become enriched and depleted in sodium chloride. A typical module of an electrodialytic salt plant has 1500 pairs of membranes, each with an effective membrane area of 1 m^2. The current density is 3.65 A/dm^2 at 620 V with a membrane spacing of about 0.75 mm. A brine concentrate containing about 118 g/l of chloride can be attained. The overall current efficiency is 73% for Na^+ and 85% for Cl^-. Typically, the membranes are divinylbenzene cross-linked polystyrene with sulphonic acid, or quaternary ammonium exchange groups; the exchange capacity is 1.8 to 2.8 meq/g at 25 °C [34].

2.1.3 Potassium

The concentration of potassium in sea water amounts to 0.38 g/l. The element is essential to plant growth and a necessary component of a balanced agricultural fertilizer. Potassium salts widely produced from brines are major constituents of the concentrated brines after sodium chloride has been removed through evaporation. Fractional crystallization usually results in recovery of potassium in the form of its chloride, primarily as carnallite $KCl \cdot MgCl_2 \cdot 6\,H_2O$. Recovery of potassium from carnallite has been described to proceed by an easy decomposition of an aqueous solution of carnallite forming crystalline potassium chloride and aqueous magnesium chloride [35].

Potassium is selectively precipitated by nitrated aromatic amines [36]. The use of dipicrylamine (hexanitrodiphenylamine) as a precipitant for potassium in sea water has been extensively studied. In comparison to others this salt is only slightly soluble in aqueous solutions (0.898 g/l at 21 °C). The magnesium, calcium, and sodium salts have a much higher solubility. The involved reaction is

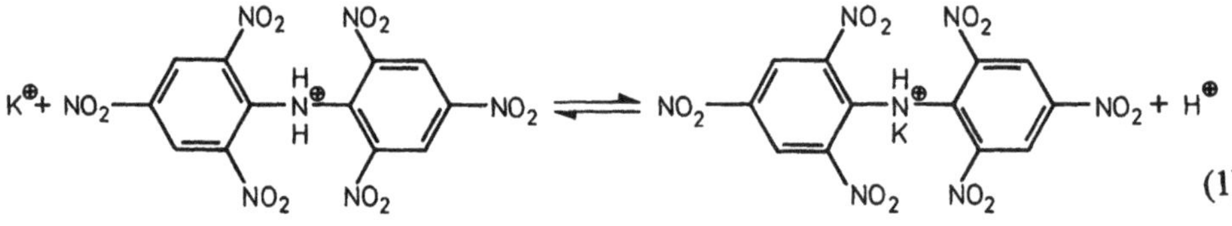

(1)

After precipitation and washing, potassium is released into solution by converting the dipicrylamine to the acid form. In batch tests with solar evaporation bitterns a potassium recovery as high as 99% was reported [37]. The insoluble acidic amine is converted into its sodium salt by treatment with NaOH and may be recycled. Dipicrylamine on activated carbon was tested for extraction of potassium from sea water [38]. Moreover, a resin was developed containing pentanitrodiphenylamine groups on a polyvinyl matrix having a reported selectivity for potassium [39].

Several inorganic ion exchangers like the zirconium salts of phosphates, silicate phosphates, molybdate phosphates, and tungstate phosphates showed selective sorption properties for potassium dissolved in sea water and brines. The potassium capacity of zirconium phosphate was found to be 25 mg K^+/g. The selectivity for potassium increased with higher drying temperatures of the exchangers. The potassium ion sorption rate exceeded that of other cations [40].

In spite of substantial research and pilot plant efforts, there has been no commercial production of potassium salts from sea water other than by fractional crystallization [5].

2.2 Alkaline Earth Metals

2.2.1 Magnesium

Magnesium is the second most abundant metal ion in sea water after sodium; about 15% of the total salt in sea water are magnesium salts. The average concentration of magnesium in sea water of 1.29 g/l is much lower than that found in commercial magnesium ores like dolomite $CaMg(CO_3)_2$, magnesite $MgCO_3$, or brucite $Mg(OH)_2$; highly pure ores are rare, however. Sea water has been a major source of magnesium compounds for about 50 years and magnesium metal has been produced from sea water since 1941. The availability and ease of processing make sea water an economically competitive source. Magnesium compounds are employed by the cement, rubber, textile, metal, chemical, and construction industries.

Commercial processes for separating magnesium from sea water start with the precipitation of $Mg(OH)_2$. In general, the alkali used is a slaked lime produced by calcining limestone or dolomite. Occasionally precipitation is carried out with sodium hydroxide, if available at a reasonable price. Basic problems associated with magnesium hydroxide precipitation are its slow settling and poor filtration characteristics. A precipitate which settles rapidly and filters easily can be achieved, for example, by recirculating the slurry or addition of organic compounds. The concentrated $Mg(OH)_2$ is washed with a counter current stream of water and filtered [5)].

Other methods of recovering magnesium from sea water have been reported, but none seem as attractive as precipitation. Among those studied are solar evaporation to produce chloride [41)], electrolysis [42)], and use of ion exchange resins with lime and carbon dioxide [43)].

A more recent ion exchange process yielding relatively pure and concentrated aqueous solutions of magnesium salts, comprises the contacting of sea water with a cation-type ion exchange resin, preferably Dowex 50 or Dowex 50W which are sulfonated copolymers of styrene cross-linked with divinylbenzene. An aqueous brine is subsequently exchanged with the magnesium-loaded resin to provide a magnesium-loaded brine which is contacted with a water-immiscible organic phase;

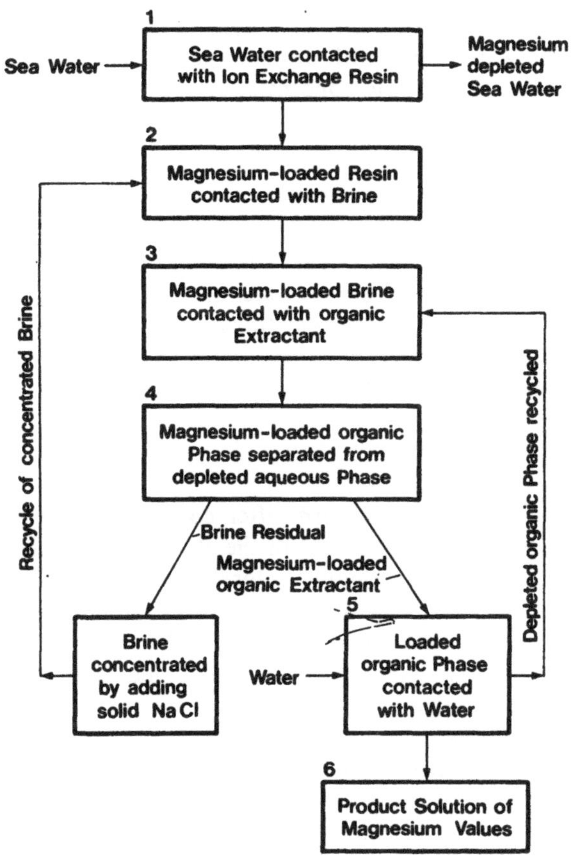

Fig. 2. Process for producing aqueous solutions of magnesium values [44)]

substantially, organic molecules with both amine and acid moieties are suitable as extractants. Magnesium salts are thereby extracted into the organic phase which is contacted with water to produce a relatively pure solution containing up to 20 wt.% of magnesium values (Fig. 2)[44].

2.2.2 Calcium

The calcium concentration in sea water amounting to 0.412 g/l is about one third that of magnesium. Commercially, crude calcium chloride is recovered along with sodium salts by the evaporation of natural brines[45] or sea water after precipitation of magnesium[46]. Other procedures are designed for the recovery of gypsum as the major calcium product[47,48]. However, the bulk of calcium is produced by mining mineral deposits like limestone, dolomite, and gypsum. Much of the recent interest in calcium separation from sea water would be more properly referred to as calcium removal, together with magnesium removal, in order to reduce scale formation problems in desalination plants[7].

2.2.3 Strontium

Although strontium is one of the major metal ion constituents in sea water, its concentration of 8.1 mg/l is about 50 times lower than that of calcium. No procedures are known for a commercial recovery of strontium from sea water. Strontium can be extracted from sea water along with uranium and other elements by hydrous titanium oxide. However, only 120-fold concentration has been reached[49,50].

2.3 Trace Metals

Most of the procedures for extracting trace metals from sea water have been developed with the objective to analytically determine these metals; till now, no trace metal including uranium can be economically recovered from sea water. An enrichment of some trace metals in marine organisms will briefly be mentioned as well.

2.3.1 Aluminum

Aluminum is extracted from sea water into chloroform after complexation with pyrocatechol violet and the formation of an ion associate of the aluminum complex with zephiramine (tetradecyldimethylbenzylammonium chloride). With 5 ml of chloroform as much as 94.7% of the aluminum was separated from 110 ml of sea water, i.e. a 20-fold aluminum concentration could easily be achieved. Several ions, such as manganese, iron(II), iron(III), cobalt, nickel, copper, zinc, cadmium, lead, and uranyl also react with pyrocatechol violet and to some extent are extracted together with aluminum. However, the interferences of these ions and other metal ions present in sea water could be eliminated by masking with sodium diethyldithiocarbamate and 8-hydroxyquinaldine. In presence of these agents all the above metal ions except aluminum were extracted into chloroform[51].

2.3.2 Titanium

Traces of titanium in sea water (at concentrations of 1 µg/l) have been preconcentrated by anion exchange from acidified samples in the presence of thiocyanate. Titanium as thiocyanate complex is strongly sorbed on a column of Amberlite CG 400 (SCN^-) and can be easily stripped by elution with 2 M hydrochloric acid containing 1.5% hydrogen peroxide. The sea water samples are favourably adjusted to 1 M HCl and 1 M NH_4SCN. Under these conditions the distribution coefficient of titanium was as high as 4000 [52].

2.3.3 Vanadium

A sorption colloid flotation method has been developed for the separation of vanadium from sea water. The separation is based on a surfactant-collector inert gas system in which vanadate is sorbed on a positively charged colloidal iron(III) hydroxide collector. The vanadate enriched collector rises to the sea water surface and floats as a separable foam with aid of sodium dodecylsulfate as surfactant and nitrogen as inert gas. The major advantages of this method are the rapid attainment of flotation and the excellent recovery of 86% vanadium based on spiked sea water samples. Flotation was found to be highly pH sensitive; optimal values were found to be 5.00 ± 0.02. In effect, at pH 4.90 a slight decline in recovery of vanadium could already be observed, whereas at pH 7 and above there was no vanadium float [53].

After chelating with 4-(2-pyridylazo)resorcinol, vanadium becomes extractable from sea water by adding a solution of tetraphenylarsonium chloride in chloroform with acetone [54].

Another concentration procedure is based on the chelation of vanadium together with manganese, copper, and zinc in sea water with ammonium pyrrolidinedithiocarbamate followed by the extraction of the chelate at pH 5.5 with chloroform [55].

The chelating ion-exchange resins Chelex-100 and Permutit S 1005 completely retain vanadium [Sc, Ni, Cu, Zn, Y, Mo, Ag, Cd, In, rare earths, W, Re (90%), Pb, Bi, and Th] from sea water. Manganese is retained quantitatively only by the Chelex resin. Vanadium (Mo, W, Re) is removed from the resins with 4 N ammonia, the other metals with 2 N mineral acids [56]. Both resins consist of a cross-linked polystyrene matrix with iminodiacetic acid [$-CH_2-N(CH_2COOH)_2$] functional groups. Alkali and alkaline earth metals do not form chelates with iminodiacetic acid except at very high pH, whereas many transition metal ions form 1:1 complexes.

Vanadium has long been recognized as a biologically active metal which has an important metabolic role in various marine organisms [53]. A vanadium concentration 5×10^5 times that found in natural sea water has been reported in the blood of the ascidian Phallusia mamillata [57]. Vanadium is concentrated in form of the blood pigment hemovanadium [58]. Moreover, vanadium enrichment factors of 4500 in scallops, 2500 in mussels, and 1500 in oysters were observed [59].

2.3.4 Chromium

Chromium (Mn, Fe, Co, Ni, Cu, Zn, Cd, Pb) can be extracted from sea water with methyl isobutyl ketone after chelating the metal ions with ammonium pyrrolidine dithiocarbamate (+8-hydroxyquinoline) and acidification to pH 3–5. The chelation is completed by heating the sea water sample to approximately 80 °C [60,61].

Chromium (Mn, Fe, Co, Ni, Cu, Zn, Cd, Pb) has also been concentrated from sea water by Chelex-100. The metals are eluted with 2.5 M HNO_3 [61].

Mussels accumulate chromium from sea water by a factor of 3.2×10^5. Concentrations of chromium up to 145 ppm have been detected in the gills of scallops (material dried at 110 °C) [59].

2.3.5 Manganese

8-hydroxyquinoline in chloroform was recommended for the extraction of manganese (Fe, Co, Ni, Cu, U) from sea water. The sample was extracted directly without pretreatment and a nearly complete separation could be achieved [62]. Including back-extraction of the manganese oxinate into 3 M HNO_3 the average recovery of manganese was better than 60% [63]. A reversed-phase liquid chromatographic technique based on a combination of chelation by 8-hydroxyquinoline with subsequent sorption on C_{18}-bonded silica gel has been used for the concentration of manganese (Fe, Co, Ni, Cu, Zn, Cd) from sea water. Enrichment factors of 50–100 are readily obtained following elution of the sorbate with methanol [64].

Manganese as well as other transition metals are separated quantitatively from alkali and alkaline earths elements in sea water with the Chelex-100 resin mentioned in chapters 2.3.3 and 2.3.4. The pH of sea water is adjusted to 5.0–5.5. Alkali and alkaline earth metal ions are eluted from the resin with ammonium acetate whereas manganese is eluted with 2.5 M HNO_3 [65].

As described for vanadium and chromium also manganese is extractable from sea water into chloroform [55] or methyl isobutyl ketone [60] after chelation with ammonium pyrrolidinedithiocarbamate.

Efficient extraction of manganese is also obtained with sodium diethyldithiocarbamate/n-butyl acetate in a pH range from 4–8. Various factors affecting extraction efficiency were recently examined in detail [66].

Of all marine organisms scallops have shown the highest enrichment factor for manganese at 5.5×10^4. The kidneys of scallops contain the remarkable quantity of 2660 ppm of Mn [59].

2.3.6 Iron

Iron(II) in sea water forms complexes with 2,4,6-tripyridyl-sym-triazine or 4,7-diphenyl-1,10-phenanthroline which are readily extracted with propylene carbonate (4-methyl-1,3-dioxolane-2-one). This extractant is colorless, nonhygroscopic, chemically stable, and more dense than water, but slightly soluble therein. Prior to complexation the iron(III) has to be reduced to its divalent state by adding hydroxylammonium hydrochloride [67].

Iron (Co, In, Zn) is highly efficiently and rapidly obtained from sea water at pH 4–10 by a single extraction with an equal volume of 0.1 M trifluoroacetylacetone in toluene. Equilibrium is reached within 10 minutes. An 88.5% yield is obtained when the phase ratio of the extracting agent to sea water is 1:40 [68].

2-Nitroso-4-chlorophenol and rhodamine B form a ternary complex with iron(II) which is extractable from sea water with benenzene or toluene at pH 4.7. A 20-fold iron concentration can easily be achieved. Hydroxylammonium chloride is used to initially reduce iron(III) [69].

A solution of both ammonium pyrrolidinedithiocarbamate and diethyldithiocarbamate has been used for complexing iron (Co, Ni, Cu, Zn, Cd, Pb) in sea water. The metal carbamate complexes are extracted from 500 ml of sea water at ca. pH 5 into 30 ml of Freon TF (1,1,2-trichloro-1,2,2-trifluoroethane) and back-extracted into 10 ml of 0.3 M nitric acid. The main advantage of this method is the transfer of the metals to a solution in which their concentrations do not change with time [70].

2-Hydroxophenyl-(2)-azonaphthol (Hyphan) is highly selective for complexing heavy metals like iron (Ni, Cu, Zn, Pb, and U). The distribution coefficients of these elements in sea water have shown to be $> 10^4$ ml/g. If one gram of Hyphan exchanger is shaken with 200 ml of sea water, 90–100% of these metals are separated. Treatment of loaded Hyphan with 1 M HCl for 1 h quantitatively eluted the metals [71]. The separation of iron (Ni, Cu, Zn, and U) from sea water in a fluidized bed of Hyphan on bead cellulose and on polystyrene has recently been studied for a period of 8 months. Initially the metals are taken up quantitatively by Hyphan on bead cellulose, whereas about one half is fixed on Hyphan on polystyrene. The uptake of iron is illustrated in Fig. 3. The distribution coefficient of iron on Hyphan fixed on bead cellulose was found to be 7.6×10^4 ml/g [72].

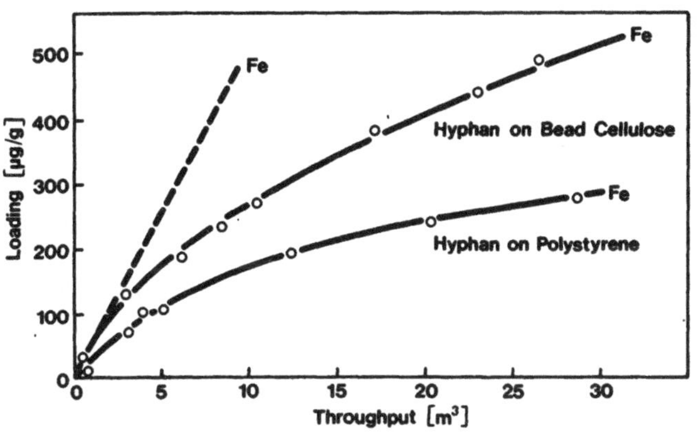

Fig. 3. Iron uptake from sea water by Hyphan on bead cellulose and on polystyrene in a fluidized bed. The dotted line indicates the loading for quantitative sorption [72]

Iron (Co, Ni) is extracted from sea water just like manganese by the systems 8-hydroxyquinoline/chloroform [62], 8-hydroxyquinoline/C_{18}-bonded silica gel [64], and ammonium pyrrolidine dithiocarbamate/methyl isobutyl ketone [61,73] or concentrated on Chelex 100 [61,65].

Scallops, oysters, and mussels accumulate iron from sea water. An enrichment factor of iron in the order of 2.9×10^5 in scallops has been reported; it is mainly concentrated in the gills which were found to contain 21600 ppm iron (material dried at 110 °C) [59].

2.3.7 Cobalt

Cobalt from acidified sea water is extracted by anion exchange on the strongly basic resin Amberlite CG 400 (SCN^-) after complexation of Co^{2+} with thiocyanate. Distribution coefficients of cobalt(II) up to 1.9×10^4 are attainable in solutions of 1 M NH_4SCN and 0.1 M HCl. Cobalt sorbed on the resin column is stripped by elution with 2 M perchloric acid [74].

At an optimal pH of 5.5–7.5 cobalt reacts with 2-nitroso-5-diethylaminophenol forming a complex which can be extracted from sea water into 1,2-dichloroethane. Some other metal ions like Fe^{3+}, Fe^{2+}, Cu^{2+}, and Ni^{2+} are also extracted into the organic phase to some extent; however, these ions, except cobalt, are stripped completely with hydrochloric acid [75].

Recently, the complexation of cobalt with 2,2′-dipyridyl-2-pyridylhydrazon at pH 5–8 and its extraction from sea water and brines into iso-amyl alcohol has been recommended. The cobalt complex can be back-extracted into dilute perchloric acid [76].

Chelex-100 retains cobalt (Cu, Pb, Ni, Zn, and Cd) from sea water. Elution of the resin with 1 M nitric acid results in a quantitative recovery of the metals [77].

Dithizone in chloroform has been proposed as an extractant for cobalt (Cd, Zn, Cu, Ni, Pb, and Ag) in sea water. Each sample was extracted twice with a 0.2% and once with a 0.02% solution of dithizone. The only treatment of the sea water before extraction is a pH adjustment to 8 for the first and to 9.5 for the second and third extraction [78].

Further extraction methods for cobalt are described in chapters 2.3.4 [61], 2.3.5 [62,64], and 2.3.6 [68].

Jellyfish and lip-fish accumulate cobalt at an enrichment factor of 2.1×10^4 [59].

2.3.8 Nickel

It is claimed that the cross-linked copolymer resin prepared from tetraethylenepentamine and toluene diisocyanate having a much greater affinity for transition metal ions than for alkaline earth ions is superior to Chelex-100 for the extraction of nickel (Cu, Zn) from sea water. At the natural pH of sea water a nearly quantitative recovery of metal ions seems to be feasible [79].

A dithizone/chloroform extraction technique for separating ng/l-levels of nickel (Cd, Cu, Zn) from sea water has been developed. The nickel extraction requires an addition of dimethylglyoxime. The metals are concentrated from 100 ml of sea water into 2.0 ml of dilute nitric acid [80].

Other methods for the extraction of nickel have already been mentioned in chapters 2.3.3 [56], 2.3.4 [61], 2.3.5 [62,64,65], 2.3.6 [70–72], and 2.3.7 [78].

Mussels concentrate nickel from sea water by an enrichment factor of 14000 [59].

2.3.9 Copper

The resins poly(malondialdehyde-triaminophenol), poly(glyoxaltriaminophenol), and poly(acrylhydroxamic acid) have been applied successfully to the extraction of copper ions from sea water in the Gulf of Naples. The Cu^{2+}-complex formed with poly-(glyoxal-triaminophenol) is shown in Fig. 4.

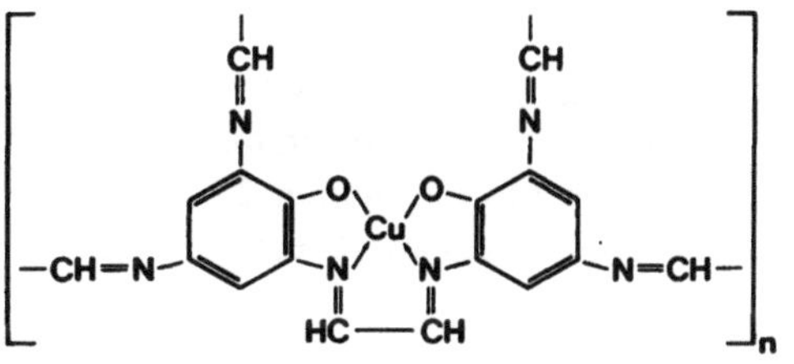

Fig. 4. Complexation of copper by poly-(glyoxal-triaminophenol) [82]

A concentration factor of 10^5 has been reported. The elution of Cu^{2+} and regeneration of the resin can be performed with dilute hydrochloric acid [81,82].

After complexation with 4-benzoyl-3-methyl-1-phenyl-5-pyrazolone the copper is extracted from sea water into methyl isobutyl ketone. 200 ml of sea water are extracted at pH with a solution of 10 ml of 0.2 wt/v % complexing agent in methyl isobutyl ketone [83].

Lead diethyldithiocarbamate dissolved in chloroform has been introduced as an effective reagent for concentrating copper (Hg, Au) from sea water. The metal ions are completely and selectively extracted from 500 ml of sea water into 10 ml of extractant solution. Lead diethyldithiocarbamate in chloroform should be added in at least 100-fold excess to the quantity of extractable metals. The sea water is favourably adjusted to pH 2–3 [84].

Copper (Pb, Zn) is obtained by passing the sea water buffered at pH 5.6 through a column of CPG-ED_3A (controlled pore glass/ethylene diamine triacetic acid). 5 l of buffered sea water are pumped through a glass tube with an inner diameter of 0.6 cm filled to a length of 10 cm at a flow rate of 5 ml/min. The metal ions are eluted with 15 ml of 1 M hydrochloric acid [85].

The retention of copper (Pb, Cd, Zn) from sea water on a column of Chelex-100 has been studied [86]. The significance of this column technique lies in its ability to separate metals without prior chemical treatment of the sample. Maximum efficiency is observed when the resin is converted to alkali or alkaline earth forms. Complete separation of inorganic or organic metal complexes dissolved in sea water can, thus, be achieved [87].

Copper and zinc have been extracted from sea water by sorption colloid flotation. The metal ions are brought to the surface in less than 5 min. using a negatively charged ferric hydroxide collector, the cationic surfactant dodecylamine, and air. 95% copper and 94% zinc could be recovered. Maximum recovery is attained at pH 7.6 [88].

Copper may also be extracted from sea water by other procedures outlined in chapters 2.3.3 [55,56], 2.3.4 [60,61], 2.3.5 [62,64,65], 2.3.6 [70–73], 2.3.7 [78], and 2.3.8 [79,80].

Oysters accumulate copper from sea water by a factor of 1.37×10^4 [59].

2.3.10 Zinc

Methods suitable for the extraction of zinc from sea water have already been mentioned in the preceding chapters in context with other trace elements. Feasible methods are the sorption colloid flotation technique (chapter 2.3.9 [88]), several column separation procedures with Chelex-100 (chapters 2.3.3 [56], 2.3.4 [61], 2.3.5 [65], 2.3.9 [86,87],

C_{18}-bonded silica gel (chapter 2.3.5 [64]), Hyphan (chapter 2.3.6 [71,72]) tetraethylenepentamine plus toluene diisocyanate resin (chapter 2.3.8 [79]), ethylene diamine triacetic acid (chapter 2.3.9 [85]), and solvent extraction techniques using the systems ammonium pyrrolidine dithiocarbamate/chloroform (chapter 2.3.3 [55]), ammonium pyrrolidine dithiocarbamate + diethyldithiocarbamate/Freon TF (chapter 2.3.6 [70]), trifluoroacetone/toluene (chapter 2.3.6 [68]), and dithizone/chloroform (chapters 2.3.7 [78] and 2.3.8 [80]).

Zinc is accumulated in gastropods, jellyfish, and sea anemones by an enrichment factor of 3.2×10^4. Scallops concentrate zinc from sea water by a factor of 2.8×10^4; the metal is preferentially accumulated in the kidneys of scallops where 2630 ppm have been found (material dried at 110 °C) [59].

2.3.11 Yttrium

Yttrium is completely retained from sea water by Chelex-100 (optimum pH 9.0) and Permutit S 1005 and eluted at 100% efficiency by means of 2 N mineral acids [56].

2.3.12 Molybdenum

Molybdenum is rapidly extracted from sea water by a sorbing colloid flotation method. Optimum collection by ferric hydroxide takes place at pH 4.0, when the colloid has an apparent maximum positive charge density and is able to sorb molybdenum nearly quantitatively as molybdate anion. From a 500 ml sample of sea water molybdenum is accumulated in the foam on the water surface in 5 min. using sodium dodecyl sulfate as surfactant and air bubbling through the solution [89,90].

Molybdenum(VI) in sea water is complexed with 4-benzoyl-3-methyl-1-phenyl-5-pyrazolone (BMPP) and extracted into isoamyl alcohol. A complete extraction is accomplished between pH 1.0–3.0: One litre of sea water acidified with 10 ml of conc. HCl, is reduced to 200 ml by evaporation. The solution is adjusted to pH 2.5 and extracted with 10 ml of 0.2 wt/v% of BMPP in isoamyl alcohol by shaking for 10 min. [91].

Chelex-100 and Permutit S 1005 are also suitable for complete extraction of molybdenum from sea water. MoO_4^{2-} is quantitatively removed from the resins by elution with 4 N ammonia [56].

Sea anemones concentrate molybdenum from sea water by a factor of 6×10^3 [59].

2.3.13 Palladium

Palladium(II), silver(I), and gold(III) are concentrated from sea water by the water-insoluble chelating agent p-dimethylaminobenzylidenerhodanine on silica gel (DMABR-SG). The chelating capacity is 11 μmol Pd, 23 μmol Ag, and 11 μmol Au per g of DMABR-SG. Palladium forms a 1:2 complex, whereas silver and gold appear to form a 1:1 complex. Quantitative separation of palladium on DMABR-SG columns required a relatively slow flow rate of 150 ml/h; in the case of silver and gold complete separation is achieved at higher flow rates of 1–2 l/h and 2–3 l/h, respectively. For the retention of palladium the pH of sea water was found to be optimal at 1.0–7.0, for silver at 1.0–6.5, and for gold at 1.0–3.5. Palladium, silver, and gold are quantitatively eluted with 0.1% thiourea in 0.1 M hydrochloric acid [92].

2.3.14 Silver

Silver ions in acidified sea water are coprecipitated with an acetone solution of 2-mercaptobenzothiazole. The precipitate is floated without surfactants by means of tiny nitrogen bubbles and collected on a sintered-glass disc. 0.05 μg of silver in 100 ml of artificial sea water have been recovered in yields higher than 95% [93].

Sorption colloid flotation has been recommended to separate silver from spiked sea water by a collector-surfactant-inert gas system consisting of lead sulfide, stearyl amine, and nitrogen. When the sea water was adjusted to pH 2, separation was found to be nearly quantitative [94].

Silver ions are completely retained from sea water by Chelex-100 and Permutit S 1005 (chapter 2.3.3 [56]). They may also be extracted with a solution of dithizone in chloroform (chapter 2.3.7 [78]) and concentrated by p-dimethylaminobenzylidene-rhodanine on silica gel (chapter 2.3.12 [92]).

Silver is accumulated in lipfish and oysters by a factor of 2.2×10^4 and 1.87×10^4, respectively [59].

2.3.15 Cadmium

The extraction of cadmium from sea water has been described in preceding chapters in context with the extraction of other heavy metals. Appropriate methods comprise the systems ammonium pyrrolidine dithiocarbamate/methyl isobutyl ketone (chapter 2.3.4 [61]) pyrrolidinedithiocarbamate plus diethyldithiocarbamate/Freon TF (chapter 2.3.6 [70]), dithizone/chloroform (chapters 2.3.7 [78] and 2.3.8 [80]), 8-hydroxyquinoline/C_{18}-bonded silical gel (chapter 2.3.5 [64]), and Chelex-100 (chapters 2.3.3 [56], 2.3.4 [61], 2.3.5 [65], 2.3.9 [86,87]).

Scallops accumulate cadmium from sea water by an exceptionally high enrichment factor of 2.26×10^6. In particular, the visceral mass dried at 110 °C contains up to 2000 ppm of cadmium [59].

2.3.16 Indium

Trifluoracetylacetone/toluene has been used for the extraction of indium (Fe, Co, Zn) from sea water, as mentioned in chapter 2.3.6 [68]. Indium is quantitatively and rapidly separated in a single extraction with an aqueous/organic volume ratio of 1:1. If this ratio is increased to 20:1 the extraction efficiency will decrease to 90%. The resins Chelex-100 and Permutit S 1005 may also be used for a complete retention of indium from sea water (refer to chapter 2.3.3) [56].

2.3.17 Rare earths, Tungsten, Rhenium

These metals can be extracted from sea water by Chelex-100 and Permutit S 1005; the rare earths and tungsten are completely retained, rhenium only with an efficiency of 90%. The rare earths are removed from the resins by means of 2 N mineral acids, tungsten and rhenium by 4 N ammonia [56].

2.3.18 Gold

The most determined attempts for recovering gold from sea water were undertaken by F. Haber [95], who after the First World War started an extensive research program with the objective of paying Germany's war debts. A procedure was developed using synthetic sea water in which gold was reduced to the metal by sodium polysulfide and removed in sulfur-coated sand filters. After the development of a suitable process, four expeditions were made on ocean-liners equipped with a special laboratory, two from Hamburg to New York, a third from Hamburg to Buenos Aires and a fourth in the North Sea. However, the results were disappointing. Haber later estimated the gold concentration in sea water to be 4 ng/l, about one-thousandth of the amount which he had expected [5].

At present the concentration of gold in sea water is considered at 10 ng/l (Table 1), approximating the value found by Haber; no regional concentration differences have been observed [1]. Taking into account the total volume of the oceans at 1.37×10^9 km^3, the total quantity of gold dissolved in sea water is calculated to be 13.7 million tons. However, the economic extraction of gold from sea water has been called an unrealisable dream [96]. Efforts to recovery gold from sea water have been reviewed earlier [7,97,98].

Polycondensation of di- and triaminothiophenol with glyoxal results in the formation of resins containing glyoxal-bis-2-mercaptoanil:

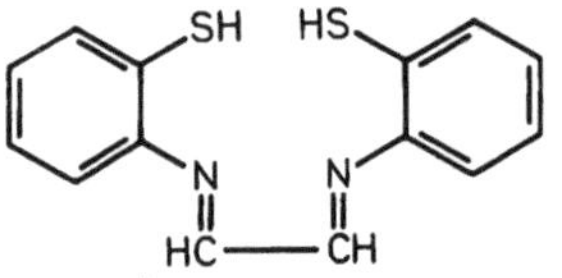

as functional groups capable of extraction gold from natural sea water. The function of the groups is explained by the favourable overlapping of the metal orbitals with the orbitals of the sulfur ligands. 1.4 μg of gold were reported to have been accumulated from 100 l of sea water in the Gulf of Naples [82].

A solution of lead diethyldithiocarbamate in chloroform was found to be suitable for the extraction of gold from sea water. The method has previously been outlined in chapter 2.3.9 for the separation of copper [84]. Moreover, gold is accumulated from sea water by p-dimethylaminobenzylidenerhodanine on silica gel as described in chapter 2.3.12 [92].

2.3.19 Mercury

Sorption colloid flotation has shown to be capable of quantitatively separating ionic mercury from sea water at levels down to 0.02 μg/l using a cadmium sulfide collector and octadecyltrimethylammonium chloride as a surfactant. The sea water samples need only to be acidified with hydrochloric acid. For flotation an adjustment to pH 1.0 is crucial. Mercury concentrations generally seemed to decrease with the depth of sea water [99].

Macroreticular polystyrene-based resins with functional aminothiazole, iminothiazole, or thiazoline groups exhibit a high selectivity for mercury(II). A thiazoline resin column has been used to concentrate mercury from sea water adjusted to pH 1 with hydrochloric acid. Maximum sorption capacity for mercury was found to be 2.8 mmole/g. The sorbed mercury is recovered quantitatively by eluting with 0.1 M HCl containing 5% thiourea [100].

Mercury can also be obtained from sea water by extraction with lead diethyldithiocarbamate/chloroform [84].

2.3.20 Lead

Lead is extractable from sea water by several procedures which are also suitable for the separation of other trace metals mentioned above. Highly selective methods are not known so far. Systems recommended for the extraction of lead are the resins Chelex-100 (chapter 2.3.3 [56], 2.3.4 [61], 2.3.5 [65], 2.3.9 [86,87], ammonium pyrrolidinedithiocarbamate/methyl isobutyl ketone (chapter 2.3.4 [60,61], pyrrolidinedithiocarbamate with diethyldithiocarbamate/Freon TF (chapter 2.3.6 [70], 2-Hydroxophenyl-(2)-azonaphtol (Hyphan) (chapter 2.3.6 [71], dithizone/chloroform (chapter 2.3.7 [78]), and ethylene diamine triacetic acid/controlled pore glass (chapter 2.3.9 [85]).

Gastropods, jellyfish, and sea anemones accumulate lead from sea water by an enrichment factor of 2600, scallops by a factor of 5300; the kidneys of scallops were shown to contain 137 ppm of lead (material dried at 110 °C) [59].

2.3.21 Bismuth, Thorium

The resins Chelex-100 and Permutit S 1005 are claimed to completely retain the heavy elements bismuth and thorium from sea water. These elements are quantitatively removed from the resins by elution with 2 N mineral acids [56].

2.4 Uranium Extraction

In view of the anticipated exhaustion of terrestrial uranium reserves in the western world at the beginning of the next century [101], the recovery of uranium from sea water has received much attention over the past three decades [102–119]. First studies on uranium extraction from sea water were carried out as early as 1953 by the Atomic Energy Research Establishment in Harwell (AERE), United Kingdom. Extensive efforts have been made in Japan since the early 1960s; from the People's Republic of China activities in this direction have been known since 1970. In the Federal Republic of Germany continuous research on the recovery of uranium from sea water has been in progress since about 1973. Investigations have also been carried out in France, Italy, Soviet Union, Finland, India, and more recently also in the United States and Sweden.

2.4.1 Uranium in Sea Water

The oceans contain about 4.5 billion tons of dissolved uranium, almost a thousandfold of the reasonably assured and estimated terrestrial uranium resources in the western world [101]. The concentration of uranium in sea water appears to be remarkably constant at about 3.3 μg/liter [120–122]. Very recent measurements of uranium concentrations in sea water samples taken in the Arctic and South Pacific Ocean down to depths of more than 5000 m confirm this mean value [123]. However, with increasing salinity of sea water a slight increase of uranium concentration is observed [124]. The molar concentration of uranium in sea water is nearly 8 orders of magnitude lower than the total concentration of the major ions [125]. Marine uranium displays no detectable deviation from the normal terrestrial U-235/U-238 isotope ratio [103,126].

Table 2. Concentration of uranium compounds dissolved in natural sea water at 25 °C at pH 8.1 [127]

Compound	Concentration [mole/l]	Fraction [wt-% U]
UO_2^{2+}	$1.53 \cdot 10^{-17}$	$0.01 \cdot 10^{-5}$
$UO_2(OH)_2$	$1.53 \cdot 10^{-12}$	0.01
$[UO_2(CO_3)_2]^{2-}$	$5.46 \cdot 10^{-11}$	0.39
$[UO_2(OH)_3]^-$	$2.43 \cdot 10^{-10}$	1.75
$[UO_2(CO_3)_3]^{4-}$	$1.37 \cdot 10^{-8}$	98.82

Uranium occurs in sea water in its highest oxidation state +6; owing to the carbonate content of sea water, uranium predominantly should exist in sea water as the tricarbonato uranylate anion $[UO_2(CO_3)_3]^{4-}$, an extremely stable complex with a formation constant of $\log \beta_3 = 22.6$. However, there is no experimental evidence for the occurrence of this complex ion in natural sea water due to its extremely low concentration. According to equilibrium constants also other uranium species are expected to occur in sea water (Table 2).

The structure of the tricarbonato uranylate complex is shown in Fig. 5.

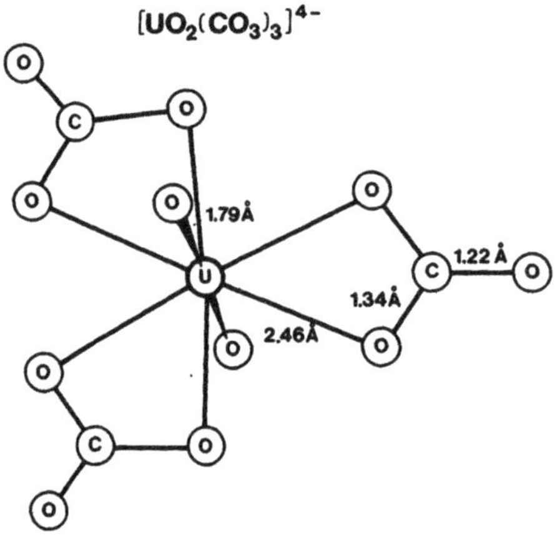

Fig. 5. Structure of $[UO_2(CO_3)]^{4-}$ [128]

The uranium atom is eight-coordinate; the linear uranyl group is equatorially surrounded by six oxygen atoms of the three bidentate carbonate groups. In the equatorial plane the radius of the complex ion amounts to 4.85 Å, thus, it is one of the largest inorganic ions existing in sea water [119].

Sea water is actually a very low grade uranium source, however, the advantage of the dissolved state and the almost inexhaustible quantities of uranium should be kept in mind. Moreover, it should be emphasized that the uranium concentration in sea water is relatively high compared to other heavy metals as for instance gold or thorium. Common metals like chromium, manganese, copper, or cobalt occur in sea water in lower molar concentrations than uranium (Table 1).

2.4.2 General Considerations and Problems

For recovering an economically significant quantity of uranium from sea water — about one ton per day — the tremendous volume of nearly 10^9 m^3 = 1 km^3 of sea water would have to be processed daily, assuming an extraction efficiency of 30%. Supplying a uranium extraction plant with 1 km^3 of fresh sea water per day already sets a problem. In order to assure a maximum available supply of uranium, the plant would continuously have to process sea water with its normal content of uranium. Therefore, the flushing time should be in the order of the plant's throughput time. Mixing of sea water forced by tides and waves may not exclude the processing of uranium depleted sea water.

A pumped system where inlet and outlet to the uranium extraction plant are widely separated or in different sea water areas, would possibly overcome the problem. However, 50 ducts of 10 m diameter would be necessary assuming a velocity of the sea water at 3 m/s within the ducts, for pumping 10^9 m^3 daily. The feasibility of such a scheme would necessarily have to be examined in terms of the energy required to pump the sea water compared to the energy potentially obtainable from the uranium if used in a light water reactor [129]. 1 ton of natural uranium yields 2 GWd of electricity in a light water reactor. At an extraction efficiency of 30%, 1 µg of uranium can be recovered from 1 l of sea water. Thus, the energy available per liter of sea water is 173 Ws_{el}, which is expended in pumping 1 liter of sea water to a height of 17.6 m. If a processing energy of this order of magnitude needs to be expended per liter of sea water, the net energy production would be zero and the recovery of uranium from sea water meaningless [107,119,129].

An ocean current of appropriate capacity should be most easily capable of supplying sea water to a uranium extraction plant. The strongest currents are found in the Gulf Stream, in the North Atlantic, and the Kuroshio in the North Pacific Ocean. The Gulf Stream transports about 10^8 m^3 of sea water per second at an average speed of 1.10 m/s; 10.6 million tons of uranium are carried along with the water per year. The average temperature of the Gulf Stream water is reported to be 24 °C in a depth of 100 m. Warm ocean currents are favourable with regard to faster sorption kinetics. The water volume transported by the Kuroshio amounts to 5×10^7 m^3 per second at a speed between 1.25–2.25 m/s; its uranium capacity is 5.3 million tons per year. The water temperature of the Kuroshio at the east coast of South Japan varies between 14–20 °C in a depth of 100 m [112].

Any process suitable for the extraction of uranium from sea water must be able to operate at the normal chemical composition of sea water for economic and ecological reasons. Any addition of reagents or removal of interfering constituents prior to the recovery of uranium must be disregarded. One of the main difficulties of uranium separation arise from the high alkali and alkaline earths concentrations in sea water. In particular, any direct or indirect acidification of sea water cannot be accepted. About 0.6 mg of hydrogen ions are necessary to neutralize 1 l of sea water or to lower its pH from 8.3 to 7. However, in a projected extraction plant which would process at least 3×10^8 m^3 of sea water per day and extract 1 ton of uranium, the same slight change of acidity would require a daily concumption of as much as ca. 9000 tons of pure sulfuric acid or 18,000 m^3 of concentrated hydrochloric acid [130].

Table 3. Uranium loading of selected inorganic, organic, and biological sorbents in natural sea water with a uranium concentration of about 3.3 µg/l; the loadings refer to the dry sorbent or to its metal content [119]

Sorbent	Functional group	Uranium loading	References
inorganic			
Hydrous aluminum oxide	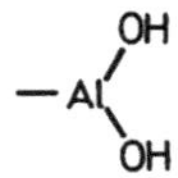	61 µg/g Al	[131]
Hydrous iron(III) oxide	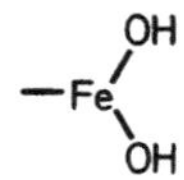	60 µg/g Fe	[131]
Silica gel	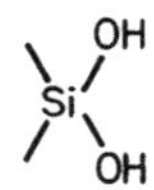	27 µg/g sorbent	[134]
Hydrous lanthanum oxide	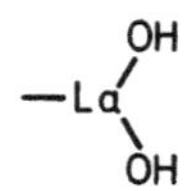	38 µg/g La	[131]
Hydrous titanium oxide	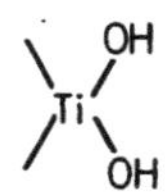	550 µg/g Ti ~200 µg/g sorbent	[103]
Hydrous titanium oxide (freshly precipitated)	—	1550 µg/g Ti	[135]
Basic zinc carbonate	—	540 µg/g Zn	[131]
Hydrous tinoxide	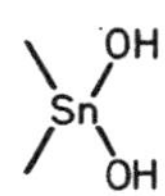	17 µg/g Sn	[131]

Hydrous zirconium oxide	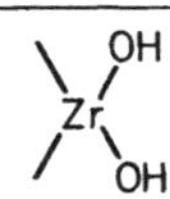	13 µg/g Zr	131)
organic			
Polystyrene-methylene-phosphonic acid	$-CH_2PO(OH)_2$	24 µg/g sorbent	131)
Resorcinol arsonic acid/ formaldehyde copolymer	$-AsO(OH)_2$	1112 µg/g sorbent	131)
Duolite ES 467	$-CH_2-NH-CH_2-PO_3Na_2$	45 µg/g sorbent	130)
Duolite ES 346 Poly(acrylamidoxime)	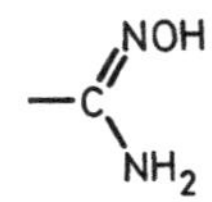	3600 µg/g sorbent	119, 125)
Oxamidoxime-terephthalic acid chloride-condensation polymer (fibers)	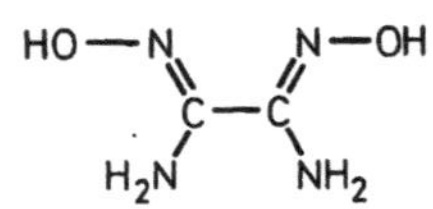	240 µg/g sorbent	136)
Poly(glyoxaltriaminophenol)	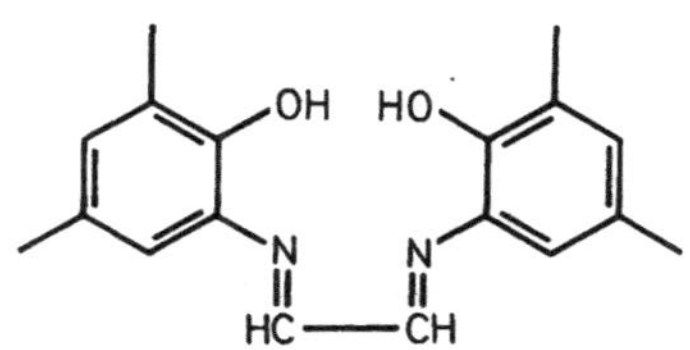	45 µg/g sorbent	134)
Hyphan on Cellulose	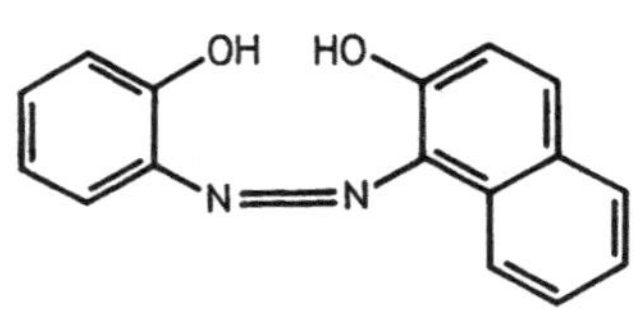	80 µg/g sorbent	137)

Table 3. (continued)

Sorbent	Functional group	Uranium loading	References
Macrocyclic hexacarboxylic acid on polystrene	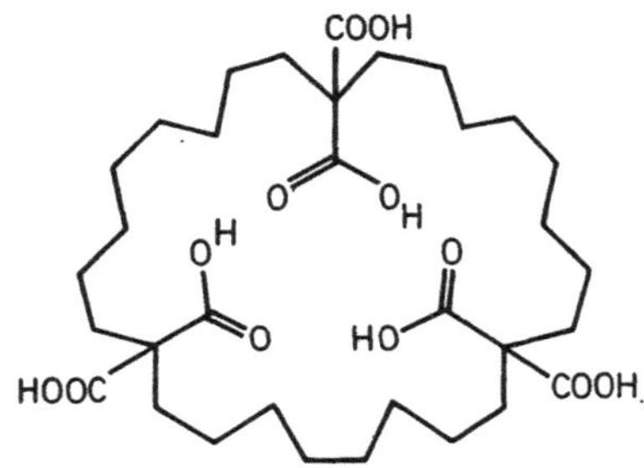	70 µg/g sorbent	[138]
Macrocycloimideresin 508	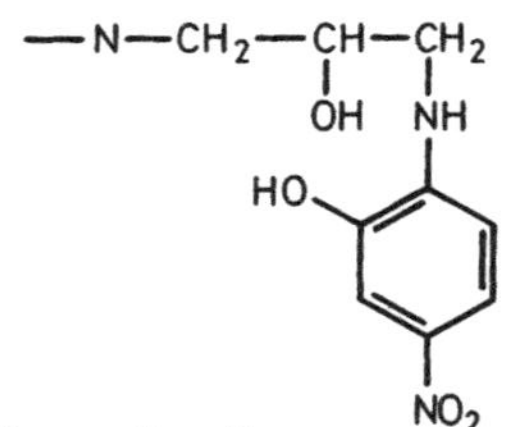	930 µg/g sorbent	[139]
Polyterephthaloyloxal-amidrazon (PTO)	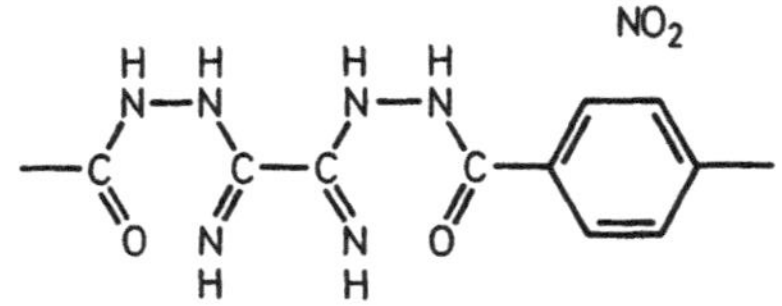	50 µg/g sorbent	[140]
biological			
Halimeda opuntia (green alga)	—	1.85 µg/g sorbent	[141]
Laurencia papillosa (red alga)	—	0.66 µg/g sorbent	[141]
Dictyota divaricata (brown alga)	—	2.14 µg/g sorbent	[141]
Oscillatoria spec. (blue-green alga)	—	2.00 µg/g sorbent	[107]
Phytoplankton (North Pacific Ocean)	—	0.86 µg/g sorbent	[142]
Zooplankton (North Pacific Ocean)	—	0.31 µg/g sorbent	[142]
Chitosanphosphate	—	2.60 µg/g sorbent	[143]

Preliminary studies on potential methods for the extraction of uranium from sea water took into consideration not only the extraction by solid sorbents, but also by solvent extraction, ion flotation, coprecipitation, and electrolysis. However, for a large-scale uranium recovery only the sorptive accumulation by use of a suitable solid sorbent seems to be feasible with regard to economic reasons and environmental impacts [119].

2.4.3 Sorbent Materials

Extensive investigations on sorbents capable of extracting uranium from natural sea water were carried out by AERE, U.K., as early as 1965 [131], some years later in Japan [132] and thereafter in the Federal Republik of Germany [133]. Numerous inorganic, organic, and biological materials have been tested, some of which are compiled in Table 3.

A suitable sorbent must be available in large quantities and at low cost. Its performance should not deteriorate in service, that is, it should be almost insoluble in sea water and eluants, and highly stable against physical, chemical, and biological degradation in order to permit long-term recycling procedures and to avoid contamination of the sea. Further, since most of the uranium is sorbed only on the surface of the sorbent particles, any loss due to attrition would mean a serious loss of uranium.

Moreover, the performance of the sorbent must be suitable for bringing it into contact with the vast volumes of sea water. Most imperative requirements pertain to the uranium uptake kinetics; the sorbent should be qualified for a rapid rate of loading in order to minimize the sorbent inventory. The quantity of required sorbent material m_s (tons of dry sorbent) is determined by the production rate P (tons of uranium per day) of a projected recovery plant, by the uranium loading b of the sorbent (ppm uranium per cycle based upon dry weight) and by the duration t (days per cycle) of a complete sorption-elution cycle:

$$m_s = P \cdot \frac{t}{b} \cdot 10^6 \qquad (2)$$

Since the elution step normally proceeds rapidly as compared to the period of sorption, the inventory of the sorbent turns out to be inversely proportional to the rate of uranium uptake [130].

The capability of any sorbent to bind uranium in sea water presupposes that the functional group of the sorbent forms a strong uranyl complex at the natural pH of sea water. The stability of such a complex should be comparable with that of $[UO_2(CO_3)_3]^{4-}$ occuring in sea water. The binding mechanism seems to proceed mostly via direct carbonate substitution by the functional group.

Lead compounds such as lead naphthalene tetracarboxylate, lead pyrophosphate, lead stannate, lead sulfide, and others which are not listed in Table 3, were proved to be effective uranium sorbents in sea water [131]. However, sorbent materials containing toxic metals must be excluded with respect to sea contamination which would be inevitable even in the case of slight solubility. On the other hand, many metal compounds are decomposed in sea water by hydrolysis and carbonate formation. Among the inorganic sorbents presented in Table 3, hydrous titanium oxide and basic zinc carbonate display remarkably high uranium loadings. However, loss of basic zinc carbonate to flowing sea water was found to be considerable [131]. Resorcinol arsonic acid/formaldehyd copolymer, the macrocycloimide-resin 508 A, the condensation polymer formed by reaction of oxamidoxime and terephthalic acid chloride, and in particular the poly(acrylamidoxime) Duolite ES 346 are worthy to note among the organic sorbents listed in Table 3. Due to hydrolysis of the carbon-arsenic bond of resorcinol arsonic acid upon contact with sea water, this polymer must be discarded. The macrocycloimide-resin and oxamidoxime have not been extensively tested to date, while poly(acrylamidoxime) has recently been screened in laboratory and test plant experiments [119,125,144–147] and is currently still under investigation. Biological materials for extraction of uranium from sea water can be excluded with regard to uranium loadings of less than 3 ppm.

Uranium-binding functional groups have been selected according to their ability to displace the carbonate ions in $[UO_2(CO_3)_3]^{4-}$ at the pH of natural sea water. The classes of compounds compiled in Table 4 proved to be effective [119,130].

Table 4. Functional groups capable of substituting carbonate ligands in $[UO_2(CO_3)_3]^{4-}$ at the pH of natural sea water

Oximes	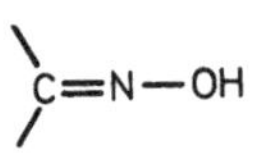	Dithiocarbaminates	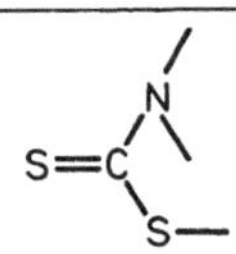
Amidoximes	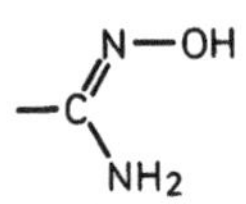	o-Dihydroxy-aromatics	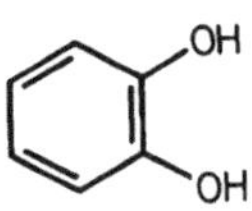
Hydroxamic acids	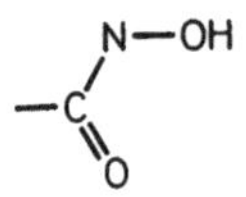	o-Diaminoaromatics	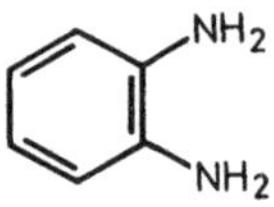
Hydroxylamines	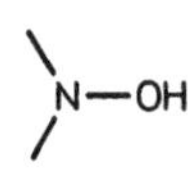	o-Hydroxyazo-aromatics	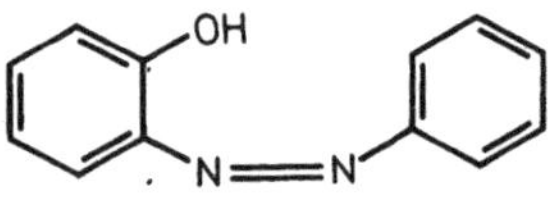
1,3-Diketones	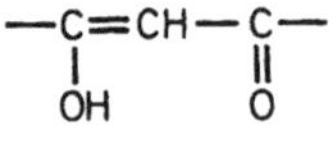		

2.4.3.1 *Hydrous Titanium Oxide*

Freshly precipitated voluminous hydrous titanium oxide having a remarkable uranium uptake of 1550 μg/g Ti (Table 3) cannot be installed in a technological process as of its lack of mechanical stability. With an increasing degree of consolidation by partial dehydration, the rate of uranium uptake as well as the uranium loading capacity of hydrous titanium oxide strongly decreases. Nevertheless, rather stable hydrous titanium oxide proves to be the most favourable sorbent among inorganic materials to date, although it does not meet the stringent requirements of sorbent properties

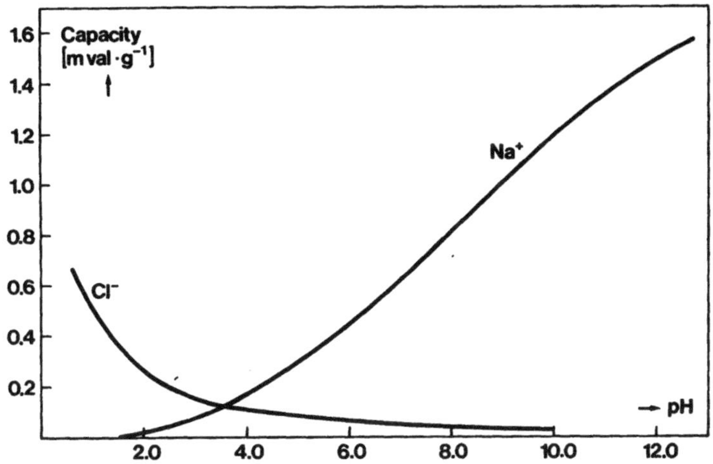

Fig. 6. Function of hydrous titanium oxide as cation and anion exchanger in dependence on pH [103]

mentioned above. The most striking disadvantages of hydrous titanium oxide are its low resistance against attrition, inadequate selectivity and uranium capacity, as well as its relatively low rate of uranium uptake.

The general method of manufacture at AERE, Harwell, U.K., where hydrous titanium oxide was first developed, is through precipitation within a solution of a titanium sulfate or chloride by addition of alkali, washing the precipitate to a controlled salt concentration, drying the precipitate to a cake, crushing the cake and sieving the particles. The product is obtained in the form of irregularly shaped glassy or opalescent granules the composition and properties of which depend on the preparative procedure. A typical composition is: 60% TiO_2, 35% H_2O, and 5% Na by weight [103].

Hydrous titanium oxide is a cation exchanger at high pH and an anion exchanger at low pH; at the normal pH 8.3 of sea water it is largely cationic (Fig. 6) [103].

The capacity of the sorbent, prepared by Japanese workers, was found to be 3 mval of Cu(II)/g [148]. Specific surfaces up to 450 m^2/g and average pore radii up to 20 Å were measured [149]. The particle density belonging to a uranium capacity of 250 μg/g sorbent, was reported to be 1.47 g/cm^3 [150]. The solubility of hydrous titanium oxide in sea water is merely about 0.1 mg/l [131]. Although uranium occurs in sea water mainly in form of the anionic complex $[UO_2(CO_3)_3]^{4-}$, it is most probably the uranyl cation which is sorbed by hydrous titanium oxide [103,151]. UO_2^{2+} sorption seems to proceed via a ligand exchange reaction tentatively described by the following Equation:

$$[UO_2(CO_3)_3]^{4-} + (\mathrm{HO\overset{|}{\underset{|}{Ti}}OH})_n \rightleftarrows [UO_2(\mathrm{O\overset{|}{\underset{|}{Ti}}O})_n]^{(2n-2)-} + 3\,HCO_3^- + (2n-3)\,H^+ \qquad (3)$$

Uranium selectivity of hydrous titanium oxide can be derived from the concentration factors of several metal ions accumulated from sea water [103] (Table 5).

Table 5. Concentration of metal ions in hydrous titanium oxide (HTO) granules after 30 days' contact with sea water [103]

Metal	Loading [g metal/kg HTO]	Concentration Factor $\left[\frac{\text{g metal/kg HTO}}{\text{g metal/l sea water}}\right]$
Na	6.11	0.58
Mg	8.63	6.8
Ca	26.3	$6.5 \cdot 10^1$
Sr	1.0	$1.2 \cdot 10^2$
Ba	0.26	$1.3 \cdot 10^4$
U	0.115	$3.5 \cdot 10^4$
Al	0.04	$4.3 \cdot 10^4$
V	0.18	$9.4 \cdot 10^4$
Cu	0.02	$1.8 \cdot 10^5$
Fe	0.26	$2.0 \cdot 10^5$
Ni	0.04	$2.2 \cdot 10^5$
Cr	0.02	$2.2 \cdot 10^5$
Mn	0.06	$3.1 \cdot 10^5$

Alkali and alkaline earth metals, except barium, are only slightly concentrated as compared with transition metals. However, total alkali and alkaline earths loadings are 300 to 400 times larger than for uranium. The uranium concentration factor turns out to be higher than that of the alkaline earth metals, but lower than the concentration factors of the other transition metals.

The uranium uptake rate of hydrous titanium oxide is demonstrated in Fig. 7.

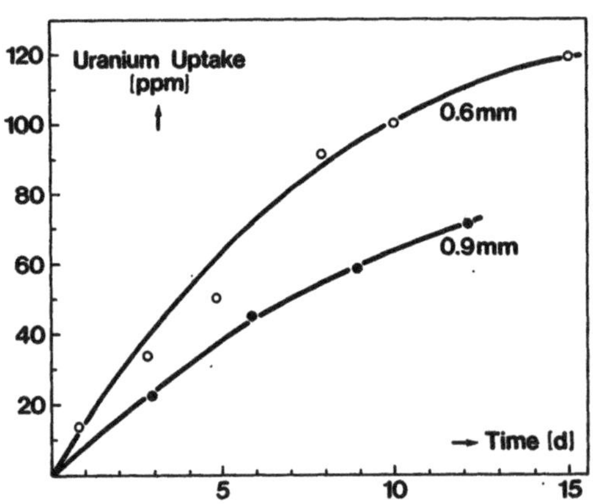

Fig. 7. Uranium uptake of hydrous titanium oxide granules 0.6 and 0.9 mm in diameter in dependence on time [152]

The uranium uptake rate increases with decreasing granular size because of the specific surface enlargement. Yet, after ten days both curves show a distinct flattening. After this time, hydrous titanium oxide granules of 0.6 mm diameter attain a uranium loading of 100 ppm. The demonstrated uptake rates seem to be realistic for mechanically rather stable hydrous titanium oxide granules.

2.4.3.2 Poly(Acrylamidoxime)

Poly(acrylamidoxime) has been tested on laboratory scale as well as in test plants in the sea in form of cross-linked, macroporous resin granules called Duolite ES 346 [153]. We succeeded in loading this resin in natural sea water with more than 3600 ppm of uranium corresponding to the uranium content of uranium ores mined today. As in the case of hydrous titanium oxide, poly(acrylamidoxime) most probably accumulates the cation UO_2^{2+} without carbonate ligands via the displacement reaction:

$$[UO_2(CO_3)_3]^{4-} + (\text{acrylamidoxime})_n \rightleftarrows [UO_2(\text{acrylamidoxime})_n]^{2+} + 3\,CO_3^{2-} \quad (4)$$

Single crystal structure analyses of uranyl acetamidoxime nitrate $[UO_2(C_2H_6N_2O)_4]$-$(NO_3)_2$ revealed the amidoxime group to act as a unidentate, non-chelating, neutral ligand. The complex contains four acetamidoximes which are attached to the uranium atom in the equatorial plane of the linear uranyl group via the oxygens of the oximes; the nitrogen atoms of the amide groups are not involved in the bonding. The proton bound to the oxygen of the oxime group is transferred to the oxime nitrogen atom during complex formation [154].

Some properties of the poly(acrylamidoxime) granules are presented in Table 6.

Table 6. Some properties of the cross-linked, macroporous poly(acrylamidoxime) resin (Duolite ES 346) [144]

Particle size[a]	0.25–0.4 mm
Water retention capacity[a]	73.2 wt-%
Bulk density[a]	0.87 g/ml
Specific gravity[a]	1.1 g/cm³
Pore volume	0.82 cm³/g
Mean pore radius	250 Å
Specific surface (BET)	150 m²/g
Uranium capacity	12.6 wt-%

[a] determined for wet granules

Table 7. Concentration of metal ions from natural sea water in poly(acrylamidoxime) (Duolite ES 346) [119]

Metal	Loading [g metal/kg resin]	Concentration Factor $\left[\frac{\text{g metal/kg resin}}{\text{g metal/l sea water}}\right]$
Na	3.92	0.36
Sr	0.087	$1.1 \cdot 10^1$
Mg	17.0	$1.3 \cdot 10^1$
Ca	13.7	$3.3 \cdot 10^1$
Ba	0.40	$2.0 \cdot 10^4$
Fe	0.41	$3.2 \cdot 10^5$
Au	0.004	$4.0 \cdot 10^5$
U	3.60	$1.1 \cdot 10^6$
V	4.34	$2.3 \cdot 10^6$
Cu	1.76	$1.8 \cdot 10^7$

The uranium loading of 3600 ppm amounts to less than 3% of the total uranium capacity. A high attrition resistance of the resin beads, expected in contrast to hydrous titanium oxide, could be demonstrated in long-term experiments. Granules exposed to natural sea water in a fluidized bed for more than 6 months showed almost no attack of the outer surface; however, the initial white color of the beads changed to brown, probably due to the complexation of transition metals. Uranium is mainly accumulated in a narrow surface layer of the beads. The high uranium selectivity of the resin can be deduced from Table 7.

The concentration of alkali and alkaline earth metals in poly(acrylamidoxime) is much lower than in hydrous titanium oxide, taking into account the 30 times higher uranium loading of the resin; the total of the alkali and alkaline earths loadings is only 10 times higher than for uranium. For vanadium and copper the remarkably high concentration factors of 2.3×10^6 and 1.8×10^7 were attained, respectively. Gold is also significantly concentrated by the resin. Yet, the total of the determined transition metal loadings turns out to be less than double of the uranium loading.

The uranium loading kinetics of the poly(acrylamidoxime) resin may be derived from Fig. 8.

During the first ten days uranium loadings were between 140–200 ppm for granular sizes between 1.0–0.2 mm in diameter; thus, the uranium loading kinetics for the resin is faster than for hydrous titanium oxide (Fig. 7). A further increase of the rate of uranium uptake could be achieved by using fibers of poly(acrylamidoxime) instead of granules; uranium loadings of 750 ppm were obtained in natural sea water after ten days.

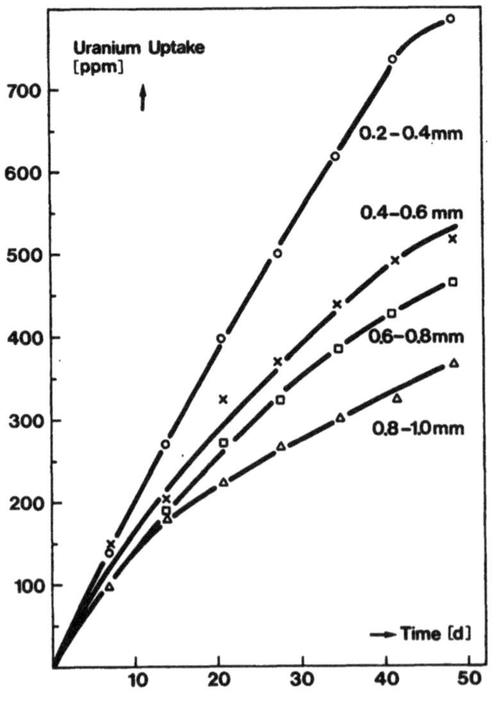

Fig. 8. Uranium uptake of poly(acrylamidoxime) granules from sea water in a fluidized bed in dependence on time and granular size [155]

2.4.4 Uranium Elution

A process for recovery of uranium from sea water comprises the regeneration of the sorbent by elution of uranium. Elution should proceed rapidly and with high yield in order to maximize the overall eficiency of the recovery process, which is measured in terms of the increase of the effective concentration, i.e. the concentration of uranium in the eluate compared to the initial concentration in the sea water. Finally, a high selectivity of the elution process is desirable [107,130]. A further concentration of uranium in the eluate up to the precipitability in the form of yellow cake can be attained by ion exchange, ion flotation, or electrodialysis.

2.4.4.1 Hydrous Titanium Oxide

Uranium is eluted from hydrous titanium oxide by 1–2 molar aqueous solutions of $(NH_4)_2CO_3$, Na_2CO_3, or $NaHCO_3$ under reformation of the tricarbonatouranylate anion according to equation (3). Integral elution efficiencies of more than 90% were obtained by using eight bed volumes of 1 M $(NH_4)_2CO_3$ for an elution period of two days [156]. Overall uranium concentrations in the eluate of about 10 ppm have been obtained; however, concentrations of more than 100 ppm are attainable in certain eluate fractions. An advantage of the carbonate elution is its uranium selectivity; on the other hand, the inevitable precipitation of alkaline earths carbonates on the hydrous titanium oxide is disadvantageous. The considerable solubility of hydrous titanium oxide in dilute mineral acids prohibits their application as eluants.

2.4.4.2 Poly(Acrylamidoxime)

Among the many potential eluants tested, only mineral acids and aqueous solutions of alkaline and ammonium carbonates yielded measurable elution efficiencies. While carbonates require higher concentrations and temperatures, dilute mineral acids turned out to be quite effective even at room temperature. Figure 8 shows six loading-elution cycles starting with a uranium content of the resin of 110 ppm.

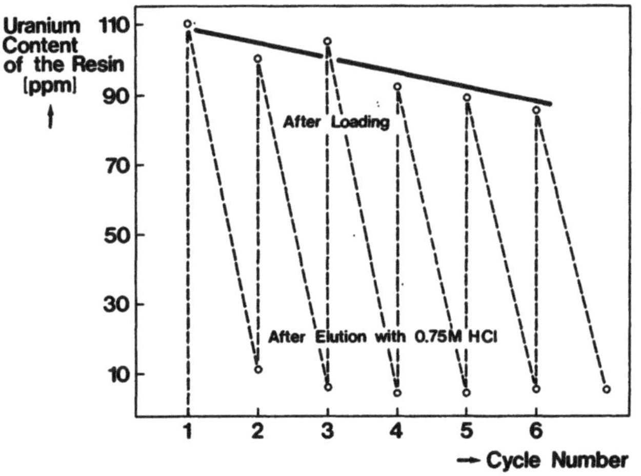

Fig. 9. Uranium loading-elution cycles of the cross-linked poly(acrylamidoxime) resin Duolite ES 346 [125]

After uranium loading in natural sea water, elution was carried out with six bed volumes of 0.75 molar hydrochloric acid resulting in an elution efficiency of about 90%. However, a serious disadvantage of the acid elution is a certain instability of the functional groups leading to a slight decrease in uranium uptake of about 5% per cycle. The attack of the functional group could clearly be established by polarographic determination of hydroxylamine formed by hydrolytic cleavage of oxime bonds.

Investigations on the stability of uranium binding functional groups revealed that isolated amidoxime groups are stable in dilute hydrochloric acid, whereas cyclic imidoximes and their derivatives proved to be unstable in this medium. Experimental studies showed the uptake of uranium from natural sea water to be closely related to the presence of cyclic imidoxime configurations formed during preparation of poly(acrylamidoxime) via reaction of poly(acrylnitrile) with hydroxylamine [119,155].

2.4.5 Recovery Processes

Several technical processes have been suggested for the recovery of uranium from sea water using solid sorbents, in particular hydrous titanium oxide. The processes have partially been tested in preliminary model plants. Generally, fixed and fluidized sorbent bed operations can be distinguished.

2.4.5.1 Fixed Sorbent Beds

First plans for the recovery of uranium from sea water on a commercial scale with hydrous titanium oxide sorbent beds and tidal movement were already made in 1964 by AERE in Harwell and the UKAEA Engineering Group in Risley, United Kingdom [102,103]. The output of the plant was estimated to be about 1000 tons of U_3O_8 per year. A design study was made for a hypothetical plant in the Menai Straits of the Irish Sea. The scheme would have involved the construction of dams and sluice gates across the northern extrance between Anglesey and Great Orme Head, and the division of the enclosed area of Beaumaris Bay into an upper and lower lagoon by means of a dam with sorbent beds according to Fig. 10.

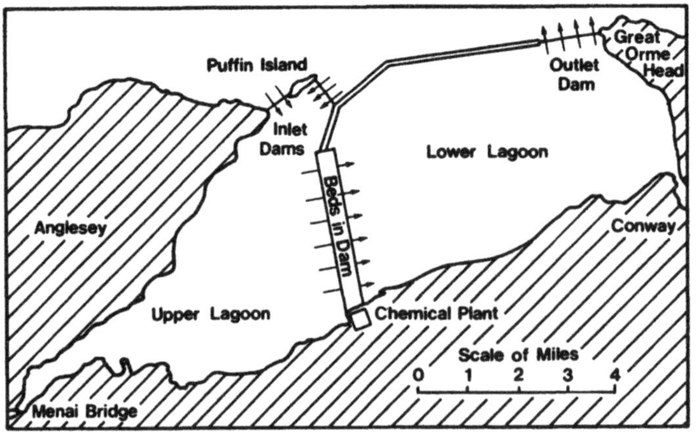

Fig. 10. Menai Straits site [103]

It was assumed that there was a sufficient north-easterly drift of water from the Irish Sea past Anglesey to maintain adequate supplies of fresh uranium-bearing water. During the rising tide the sea water flows into the upper lagoon and out of the lower lagoon during falling tide. The sorbent beds would be situated in the dam separating the upper and lower lagoons. A chemical plant for the associated chemical operations would be located at an appropriate point.

In Japan on the island of Shikoku near Nio town, Kagawa Prefecture, a pilot plant is presently under construction to recover 10 kg of uranium per year from the water of the Seto inland sea. 1500 m^3 of sea water will be led into a pool and pumped into 10 columns equipped with fixed beds of hydrous titanium oxide. Desorption of uranium by carbonate solutions will produce an eluate containing 10 ppm of uranium. Both ion-exchange membrane and ion flotation methods will be incorporated in the pilot facility to concentrate the uranium to about 2800 ppm. This project of the Metal Mining Agency of Japan (MMAJ) which is supervised by the Agency of Natural Resources and Energy, will be joined by Tokuyama Soda Co., Asahi Chemical Industry Co., and Mitsubishi Metal Corp. The project fits into a long-term scenario written by the Council for Ocean Development. The council envisaged, first, a pilot plant construction and operation in the 1980s to confirm feasibilities, second, a semicommercial plant construction with recovery capacity up to tens of tons of uranium per year by 1990, and third, a fully commercial plant with annual capacity of about 1000 tons/year [157].

Very recently an interesting concept has been designed in Sweden, utilizing the wave energy for supplying a uranium recovery plant with sea water as well as creating a pressure head which can force the sea water through the fixed beds of a suitable sorbent [158].

Floating, anchored units of about 400 m in length bear sea water reservoirs at a level which is higher than the water level in the sea. Waves will rush up on a sloping plane and fill the reservoirs with sea water (Fig. 11).

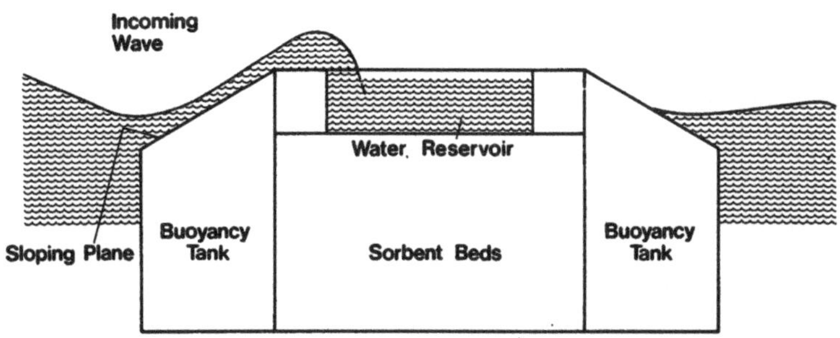

Fig. 11. Schematic layout of a floating unit utilizing wave energy for recovery of uranium from sea water [158]

More than 20 such units are required for the extraction of 600 t of uranium per year. A sea current in the area of the recovery plant should provide for the processing of uranium bearing fresh sea water.

2.4.5.2 Fluidized Sorbent Beds

Fluidized beds, in which the sorbent granules are kept in suspension in a flow of sea water or eluant, have some advantages as compared with fixed sorbent beds. Besides the more effective contact of the sorbent with the liquid, the fluidized bed generally permits larger throughputs of sea water, since there is less danger of bed blockage by suspended materials. A disadvantage of the fluidized bed may be a stronger attrition of the sorbent particles.

The first definite concept of a uranium extraction plant based on a fluidized bed was suggested in Germany [106,159]. A key role in this concept is played by the sorption module shown in Fig. 12.

The sea water flows through the diffusor (1) into the settling chamber (2) where the kinetic energy of the oncoming ocean current is converted into a static pressure gain, which is used to form the fluidized bed (3) by suspending the sorbent particles between the perforated plates (4). An ocean current velocity of 1.5 m/sec results in a water-column pressure head of more than 10 cm. The uranium depleted sea water leaves the fluidized bed through the upper perforated plate and is carried off by the sea current. Several of these sorption modules may be arranged in such a way that the fluidized bed of the preceding module is placed on the diffusor of the subsequent module forming a huge wash-board like sorbent bed area. A concept of a plant for the recovery of about 100 toms of uranium from sea water is sketched in Fig. 13.

Sorption modules described in Fig. 12 form the outriggers of the semidiving ship anchored in a warm sea current in a depth of about 40 m to minimize surface wave effects. The outriggers on both sides of the central body have an width of about 150 m,

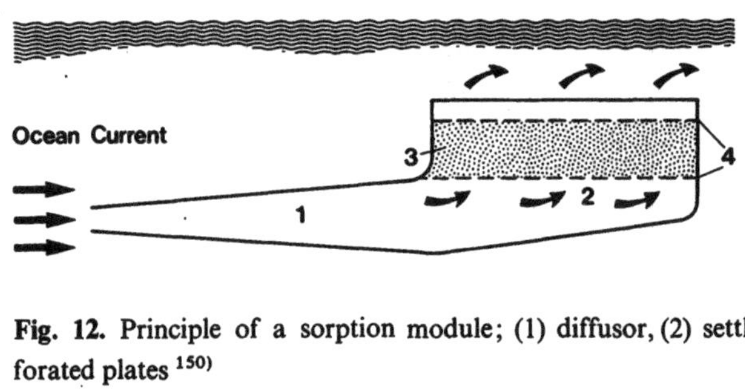

Fig. 12. Principle of a sorption module; (1) diffusor, (2) settling chamber, (3) fluidized bed, (4) perforated plates [150]

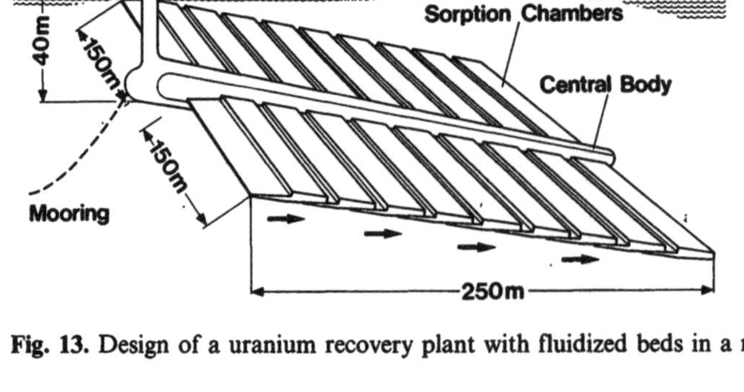

Fig. 13. Design of a uranium recovery plant with fluidized beds in a natural sea current [160]

Table 8. Extracting Systems used for Metal Separation from Sea water

Extracting Systems	Metal Ions	Remarks	References
Ion-exchangers			
Chelex -100 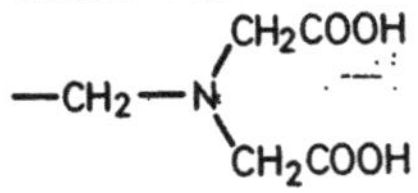	Sc, V, Cr, Mn, Fe, Co, Ni, Cu, Zn, Y, Mo, Ag, Cd, In, Rare earths, W, Re, Pb, Bi, Th	Cation elution with dilute mineral acids, anion elution with 4 N ammonia	56, 61, 65, 86, 87)
Retardant 11-A8 —CH_2COOH —$CH_2N^+(CH_3)_3Cl^-$	Li	Concentration factor of 30 in the ethyl alcohol eluant	16)
Amberlite CG 400 —$CH_2N^+(CH_3)_3SCN^-$	Ti, Co	Sorption after complexation with SCN^-. Elution with 2 M mineral acids	52, 74)
Poly(tetraethylenepentamine)-polyurea resin	Ni, Cu, Zn	Nearly quantitative recovery at natural pH of sea water	79)
Poly(glyoxaltriaminophenol) 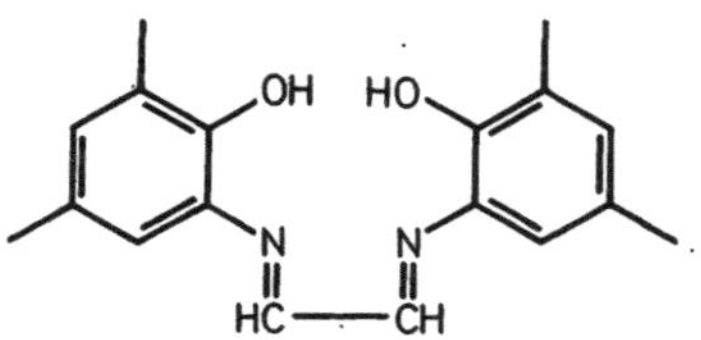	Cu, U	Separation at natural pH of sea water. Elution with dilute mineral acids	81, 82)
Poly(glyoxaltriaminothiophenol) 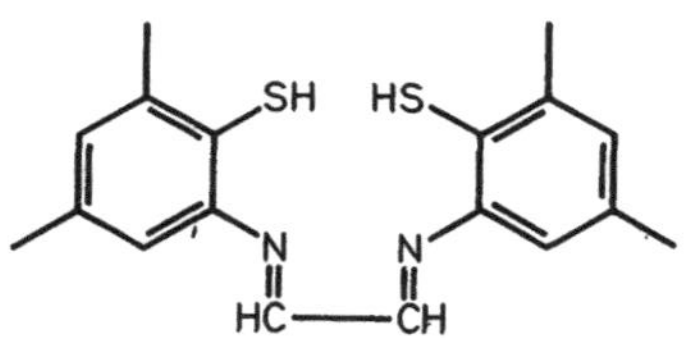	Au	Separation at natural pH of sea water. Elution with dilute mineral acids	82)

Table 8. (continued)

Extracting Systems	Metal Ions	Remarks	References
Hydrous aluminum oxide 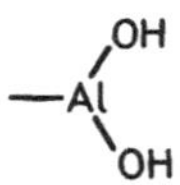	Li	Elution with boiling water. Maximal concentration factor 46 in the eluate	17–21)
Dipicrylamine 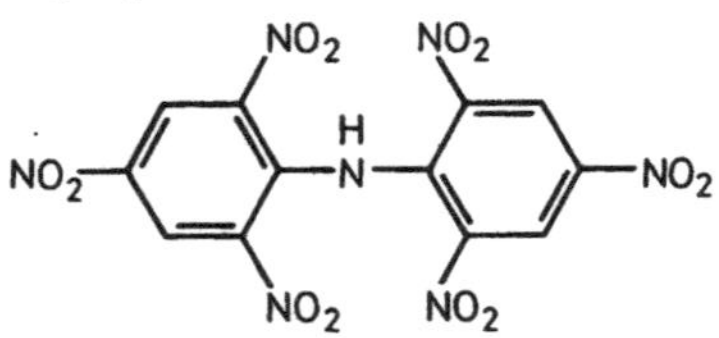	K	Precipitation of the potassium salt. Elution of K^+ with mineral acids	36–39)
Zirconium phosphates	K	Potassium capacity 25 mg/g	40)
Dowex 50 $—SO_3^-H^+$	Mg	See Figure 2!	44)
Hydrous titanium oxide 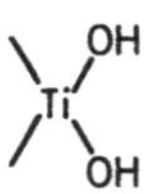	Na, Mg, Al, Ca, V, Cr, Mn, Fe, Ni, Cu, Sr, Ba, U	Sorption at natural pH of sea water. Elution with dilute mineral acids. Selective elution of U with 2 M $(NH_4)_2CO_3$	49, 50)
Hyphan on cellulose or polystyrene 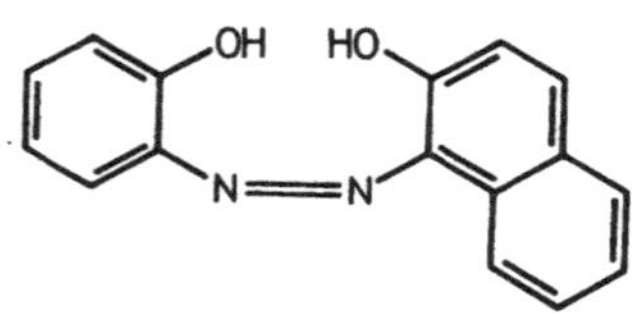	Fe, Ni, Cu, Zn, Pb, U	Almost complete separation. Elution with 1 M HCl	71, 72)

Thiazoline on polystyrene	Hg	Extraction at pH 1. Complete elution with 0.1 M HCl containing 5% thiourea	[100]
8-hydroxyquinoline on C_{18}-bonded silica gel	Mn, Fe, Co, Ni, Cu, Zn, Cd	Concentration factors 50–100 in the methanol eluate	[64]
Ethylene diamine triacetic acid on pore glass	Cu, Zn, Pb	Sea water buffered at pH 5.6. Elution with 1 M HCl	[85]
p-Dimethylaminobenzylidenerhodanine on silica gel	Pd, Ag, Au	Quantitative retention in acidified sea water. Elution with 0.1 M HCl containing thiourea	[92]
Chitin	Co, Sb, Au, Hg	Sorption at pH 7	[161]
Chitosan	Co, Zn, Cu, Mo, Pd, Sb, Cs, Ir, Au, Hg, U	Sorption at pH 7	[161]
Duolite ES 346 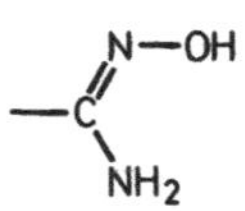	Na, Mg, Ca, V, Fe, Cu, Sr, Ba, Au, U	Sorption at natural pH of sea water. Elution with dilute mineral acids. Uranium capacity >3600 ppm	[119, 125, 144]
Flotation systems:			
Iron(III)hydroxide + Sodium dodecylsulfate	V, Mo, U	More than 80% recovery at the appropriate pH	[53, 89, 90]
Iron(III)hydroxide + Dodecylamine	Cu, Zn	More than 90% recovery at pH 7.6	[88, 90]
Lead sulfide + Stearyl amine	Ag	Almost quantitative separation at pH 2	[94]
2-Mercaptobenzo-thiazole coprecipitate without surfactant	Ag	More than 95% recovery at pH 1	[93]
Cadmium sulfide + Octadecyltrimethylammonium chloride	Hg	Quantitative separation at pH 1	[99]
Complexing agents/organic solvents:			
Ammonium pyrrolidine dithiocarbamate/methyl isobutyl ketone, chloroform or Freon TF	V, Cr, Mn, Fe, Co, Ni, Cu, Zn, Cd, Pb	Extraction at pH 3–5	[55, 60, 61, 70, 73]
4-Benzoyl-3-methyl-1-phenyl-5-pyrazolone/iso-amyl alcohol or methyl isobutyl ketone	Cu, Mo	Extraction of Cu at pH 7, Mo extraction at pH 1–3	[83, 91]

Table 8. (continued)

Extracting Systems	Metal Ions	Remarks	References
Dithizone/chloroform	Co, Ni, Cu, Zn, Ag, Cd, Pb	Extraction at pH 8–9.5	78, 80)
8-Hydroxyquinoline/chloroform	Mn, Fe, Co, Ni, Cu, U	Extraction of Mn at pH 2. No pretreatment of sea water for the extraction of the other metals	62, 63)
Trifluoroacetylacetone/toluene	Fe, Co, Zn, In	Complete extraction of Co and Zn requires the addition of isobutylamine	68)
2-Nitroso-4-chlorophenol + rhodamine B/toluene	Fe	Extraction at pH 4.7 after reduction of Fe(III) to Fe(II)	69)
2,4,6-Tripyridyl-sym-triazine/propylene carbonate	Fe	Extraction at pH 2.5–8 after Fe(III) reduction	67)
2,2′-Dipyridyl-2-pyridylhydrazone/iso-amyl alcohol	Co	Extraction at pH 5–8	76)
2-Nitroso-5-diethylaminophenol/1,2-dichloroethane	Co	Extraction at pH 5.5–7.5	75)
4-(2-Pyridylazo)resorcinol/tetraphenylarsonium chloride in chloroform + acetone	V	Addition of tartrate- and acetate-buffer to sea water	54)
Sodium diethyldithiocarbamate/n-butyl acetate	Mn	Extraction at pH 4–8	66)
Lead diethyldithiocarbamate/chloroform	Cu, Au, Hg	Extraction at pH 2–3	84)
Pyrocatechol + zephiramine/chloroform	Al	Interfering metal ions are extracted with sodium diethyldithiocarbamate and 8-hydroxyquinaldine	51)

the total length of the ship is about 250 m. The central body contains the elution plant. The fluidized sorbent behaves like a fluid with increased viscosity, allowing the sorbent granules to migrate from the feeding to the discharge station [150].

3 Conclusion

The extracting systems which have been used for the separation of metal ions from sea water are compiled in Table 8.

Numerous procedures recommending ion-exchangers, flotation systems, or solvent extractions of the metals have been developed predominantly for the preconcentration and analytical determination of metal ions in sea water. Methods for the recovery of metals from sea water are mainly restricted to lithium, sodium, potassium, magnesium, calcium, gold, and uranium. In view of the similar behaviour of many transition metals towards complexing agents, the procedures often would require a higher degree of selectivity; high concentration factors of trace metals in extracting systems can only be expected, if the bonding of alkali and alkaline earth metals is weaker by several orders of magnitude. Extensive investigations are required particularly for clarifying the composition and structure of the metal complexes formed by different extracting agents.

Among the trace metals which have been taken into consideration for recovery from sea water, notably uranium has received special attention. The extraction of uranium from sea water at an economically significant production rate of about 1 ton per day involve multiple problems which, however, could possibly be solved in the future. Whether hydrous titanium oxide or poly(acrylamidoxime), recently shown to have a superior uranium capacity, uptake rate, and mechanical stability, but is not completely regenerable, will be installed in a projected recovery plant, may depend on the special kind of process. The application of fluidized sorbent beds using the kinetic energy of strong and warm ocean currents, seems to be advantageous, even if fixed sorbent beds and pumping of sea water may be technically realized more readily.

Obviously, the entire concept of uranium recovery from sea water is invalid if the recovery plant can not deliver a substantial energy gain. However, energy balance considerations as well as cost analyses are less significant without definite concepts for process engineering. Preliminary cost estimates based on supposed technological processes claim the total costs for the recovery of uranium from sea water to be some 100 \$/lb of U_3O_8, which is roughly one order of magnitude higher than the present spot price of 23 \$/lb U_3O_8 of terrestrial uranium [165]. However, it should be emphasized that such an increase of the uranium price would have very little effect on the specific electricity price. Nevertheless, it has been pointed out that uranium from sea water will not find any customer as long as less expensive uranium is available [166].

4 References

1. Brewer, P. G.: Chemical Oceanography (J. P. Riley, G. Skirrow, Eds.) 2nd ed., Vol. 1, p. 415, Academic Press 1975
2. Burton, J. D.: Chem. and Ind. *1977*, 550

3. Förstner, U., Wittmann, G. T. W.: Metal Pollution in the Equatic Environment, 2nd Ed., Springer Verlag 1981
4. Wilson, T. R. S.: Chemical Oceanography (J. P. Riley, G. Skirrow, Eds.) 2nd ed., Vol. 1, p. 365, Academic Press 1975
5. McIlhenny, W. F.: Chemical Oceanography (J. P. Riley, G. Skirrow, Eds.) 2nd ed., Vol. 4, p. 155, Academic Press 1975
6. McIlhenny, W. F., Ballard, D. A.: Desalination and Ocean Technology (S. Levine, ed.) Dover Publications, New York 1963
7. Tallmadge, J. A., Butt, J. B., Solomon, H. J.: Ind. Eng. Chem. *56*, 44 (1964)
8. McIlhenny, W. F.: Encyclopedia of Chemical Technology (A. Standen, ed.), 2nd Ed., Vol. *14*, 150, Interscience, New York 1967
9. Hanson, C., Murthy, S. L. N.: The Chemical Engineer *1972*, 295
10. Setharam, B., Srinivason, D.: Chemical Engineering World *13*, 63 (1978)
11. Massie, K. S.: Elsevier Oceanography Series, Vol. *24b*, 569 (1980)
12. Ogata, N.: Nippon Kaisui Gakkaishi *35*, 134 (1981)
13. Christensen, J. J., McIlhenny, W. F., Muehlberg, P. E., Smith, H. G.: Office of Saline Water Research and Development, Progress Rep. No. 245, Supt. of Documents, Washington, D.C., 1967
14. Steinberg, M., Dany, V.-D.: Geological survey professional paper *1005*, 79 (1976)
15. Takeuchi, T.: J. Nucl. Sci. Technol. *17*, 922 (1980)
16. Koyanaka, Y., Yasuda, Y.: Suiyo Kaishi *18*, 523 (1977)
17. Kitamura, T., Wada, H.: Jpn. Pat. 956532 (1978)
18. Kitamura, T., Wada, H.: Bull. Soc. Sea Water Sci. Jpn. *32*, 78 (1978)
19. Kitamura, T., Wada, H., Ooi, K., Takagi, N., Katoh, S., Fujii, A.: Rep. Gov. Ind. Res. Inst. Shikoku *12*, 1 (1980)
20. Ooi, K., Wada, H., Kitamura, T., Katoh, S., Sugasaka, K.: ibid. *12*, 6 (1980)
21. Wada, H., Kitamura, T., Fujii, A., Katoh, S.: The Chem. Soc. of Japan *1982*, 1156
22. Aikawa, H., Kato, Y.: Japan. Patents 132,465 to 132,467 (1939)
23. Cady, W. R., Julien, A. P., Saunders, D. J.: U.S. Pat. 2,764, 472 (1956)
24. Kane, Y. P., Kamat, B. K.: India Pat. 46,740 (1953)
25. Wiseman, J. V.: U.S. Pat. 2,784,056 (1957)
26. Nakao, S., et al.: Japan. Pat. 2267 (1952)
27. Kume, T.: Records Oceanog. Works Japan *12*, 57 (1957)
28. Imamura, M., Izawa, S.: Japan. Pat. 8774 (1956)
29. Inoue, S.: Japan. Pat. 175,522 (1948)
30. Nakazawa, H., Atsugi, T., Onoe, K.: Japan. Pat. 4026 (1955)
31. Nakazawa, H.: Japan. Pat. 2615 (1955)
32. Yamamura, T., Nomiyama, Y.: Japan. Pat. 181,089 (1949)
33. Sueda, H., Nakahara, S., Yamamura, T.: Japan. Pat. 175,043 (1948)
34. Tsunoda, Y.: Proc. Int. Symp. Water Desalination 1st, Vol. 3, p. 225 U.S. Dept. of Interior, Washington, D.C. 1967
35. Otaya, H., Shibata, T., Myojo, H.: Ann. Rept. Shionogi Res. Lab. No. 2, 116 (1952)
36. Kielland, J.: Ger. Patents 691,366 (1940); 704,545; 704,546; 715,199; 715,200 (1941)
37. Massazza, F., Riva, B.: Ann. Chim. (Rome) *48*, 664 (1958)
38. Isobe, H., Shimamoto, T.: Japan. Pat. 2271 (1959)
39. Skogseid, A.: Norwegian Pat. *83*, 579 (1954)
40. Matsushita, H., Takayanagi, T.: Nippon Kaisui Gakkai-Shi *25*, 269 (1972)
41. Piromallo, A.: Ital. Pat. 460,207 (1950)
42. Schacher, O.: Bull. Res. Council Israel *5C*, 100 (1955)
43. Akabori, S., et al.: Japan. Pat. 179,562 (1949)
44. Baumann, W. C.: U.S. Pat. 3,615,181 (1971)
45. Staab, W. A.: The Compass *24*, 5 (1946)
46. Trauffer, W. E.: Pit and Quarry *30*, 43 (1938)
47. Munakata, E., Suzuki, A.: Japan. Pat. 7,566 (1956)
48. Myers, C. B.: Brit. Pat. 755,948 (1956)
49. Schwochau, K., Astheimer, L., Schenk, H. J., Witte, E. G.: Chemiker Ztg. *107*, 177 (1983)
50. Keen, N. J.: J. Brit. Nucl. Energy Soc. *7*, 178 (1968)

51. Korenaga, T., Motomizu, S., Tôei, K.: Analyst *105*, 328 (1980)
52. Kiriyama, T., Haraguchi, M., Kuroda, R.: Fresenius Z. Anal. Chem. *307*, 352 (1981)
53. Hagadone, M., Zeitlin, H.: Anal. Chim. Acta *86*, 289 (1976)
54. Monien, H., Stangel, R.: Fresenius Z. Anal. Chem. *311*, 209 (1982)
55. Kusaka, Y., Tsuji, H., Tamari, Y., Sagawa, T., Ohmori, S., Imai, S., Ozaki, T.: Radional. Chem. *37*, 917 (1977)
56. Riley, J. P., Taylor, D.: Anal. Chim. Acta *40*, 479 (1968)
57. Bielig, H. J., Bayer, E., Califano, L., Wirth, L.: Pubbl. Staz. Zool. Napoli *25*, 26 (1954)
58. Bayer, E.: Experientia *12*, 365 (1956)
59. Horne, R. A.: Marine Chemistry, Wiley-Interscience, New York, 1969
60. Kubo, Y., Nakazawa, N., Sato, M.: Sekiyu Gakkai Shi *16*, 588 (1973)
61. Sturgeon, R. E., Berman, S. S., Desaulniers, A., Russel, D. S.: Talanta *27*, 85 (1980)
62. Armitage, B., Zeitlin, H.: Anal. Chim. Acta *53*, 47 (1971)
63. Klinkhammer, G. P.: Anal. Chem. *52*, 117 (1980)
64. Sturgeon, R. E., Berman, S. S., Willie, S. N.: Talanta *29*, 167 (1982)
65. Kingstone, H. M., Barnes, I. L., Brady, T. J., Rains, T. C., Champ, M. A.: Anal. Chem. *50*, 2064 (1978)
66. Roberts, R. F.: Univ. of South Florida, Dissertation (1981)
67. Stephens, B. G., Suddeth, H. A.: Anal. Chem. *39*, 1478 (1967)
68. Lee, M.-L., Burrell, D. C.: Anal. Chim. Acta *62*, 153 (1972)
69. Korenaga, T., Motomizu, S., Tôei, K.: Talanta *21*, 645 (1974)
70. Danielsson, L.-G., Magnusson, B., Westerlund, S.: Anal. Chim. Acta *98*, 47 (1978)
71. Lieser, K. H., Gleitsmann, B.: Fresenius Z. Anal. Chem. *313*, 203 (1982)
72. Lieser, K. H., Gleitsmann, B.: ibid. *313* 289 (1982)
73. Kremling, K., Petersen, H.: Anal. Chim. Acta *70*, 35 (1974)
74. Kiriyama, T., Kuroda, R.: Fresenius Z. Anal. Chem. *288*, 354 (1977)
75. Motomizu, S.: Anal. Chim. Acta *64*, 217 (1973)
76. Kouimtzis, T. A., Apostolopoulou, C., Staphilakis, I.: ibid. *113*, 185 (1980)
77. Lamathe, J.: ibid. *104*, 307 (1979)
78. Ármannsson, H.: ibid. *110*, 21 (1979)
79. Leyden, D. E., Patterson, T. A.: Anal. Chem. *47*, 733 (1975)
80. Smith, R. G., Windom, H. L.: Anal. Chim. Acta *113*, 39 (1980)
81. Bayer, E., Fiedler, H.: Angew. Chem. *72*, 921 (1960)
82. Bayer, E., Fiedler, H., Hock, K.-L., Otterbach, D., Schenk, G., Voelter, W.: ibid. *76*, 76 (1964)
83. Akama, Y., Nakai, T., Kawamura, G.: Nippon Kaisui Gakkaishi *33*, 120 (1979)
84. Lo, J. M., Wei, J. C., Yeh, S. J.: Anal. Chem. *49*, 1146 (1977)
85. Guedes da Mota, M. M., Jonker, M. A., Griepink, B.: Fresenius Z. Anal. Chem. *296*, 345 (1979)
86. Florence, T. M., Batley, G. E.: Talanta *22*, 201 (1975)
87. Abdullah, M. I., El-Rayis, O. A,, Riley, J. P.: Anal. Chim. Acta *84*, 363 (1976)
88. Kim, Y. S., Zeitlin, H.: Separ. Sci. *7*, 1 (1972)
89. Kim, Y. S., Zeitlin, H.: ibid. *6*, 505 (1971)
90. Kim, Y. S., Zeitlin, H.: Chem. Commun. *1971*, 672
91. Akamu, Y., Nakai, T., Kawamura, F.: Nippon Kaisui Gakkaishi *33*, 180 (1979)
92. Terada, K., Morimoto, K., Kiba, T.: Anal. Chim. Acta *116*, 127 (1980)
93. Hiraide, M., Mizuike, A.: Bull. Chem. Soc. Japan *48*, 3753 (1975)
94. Rothstein, N., Zeitlin, H.: Anal. Letters *9*, 461 (1976)
95. Haber, F.: Z. Angew. Chem. *40*, 303 (1927)
96. Lancaster, F. H.: Gold Bulletin (Johannesburg) *6*, 111 (1973)
97. Purkayastha, B. C., Ranjandas, N.: Science and Culture (Calcutta) *31*, 403 (1965)
98. Mero, J. L.: The Mineral Resources of the Sea, Elsevier Publishing Company, New York, 1965
99. Voyce, D., Zeitlin, H.: Anal. Chim. Acta *69*, 27 (1974)
100. Sugil, A., Ogawa, N., Hashizume, H.: Talanta *27*, 627 (1980)
101. Uranium Resources, Production and Demand. A Joint Report by the OECD Nuclear Energy Agency and the Internat. Atomic Energy Agency, Febr. 1982

102. Davies, R. V., Kennedy, J., McIlroy, R. W., Spence, R., Hill, K. M.: Nature *203*, 1110 (1964)
103. Kenn, N. J.: J. Brit. Nucl. Energy Soc. *7*, 178 (1968)
104. Ogata, N.: Genshiryoku Kogyo *16*, 19 (1970)
105. Llewelyn, Y. I. W.: Atom (London *238*, 214 (1976)
106. Bals, H. G.: Rep. BMFT-FB (UR 1366) (1976)
107. Schwochau, K., Astheimer, L., Schenk, H.-J., Schmitz, J.: Rep. Jülich-1415 (1977)
108. Kanno, M.: J. Atomic Energy Soc. Japan *19*, 586 (1977)
109. Schwochau, K.: Nachr. Chem. Techn. Lab. *27*, 563 (1979)
110. Schwochau, K., Astheimer, L., Schenk, H.-J., Witte, E. G.: Chem.-Ing.-Techn. *51*, A 706 (1979)
111. Witte, E. G., Astheimer, L., Schenk, H.-J., Schwochau, K.: Ber. Bunsenges. Phy. Chem. *83*, 1121 (1979)
112. Rodman, M. R., Gordon, L. I., Chen, A. C. T., Campbell, M. H., Binney, S. E.: Rep. XN-RT-14 (1979)
113. Campbell, M. H., Frame, J. M., Dudey, N. D., Kiel, G. R., Mesec, V., Woodfield, F. M., Binney, S. E., Jante, M. R., Anderson, R. C., Clark, G. T.: Rep. XN-RT-15 (1979)
114. Sugasaka, K.: Kagaku Gijutsushi Mol *18*, 25 (1980)
115. Kanno, M., Saito, K.: NEUT Research Report 80-08 (1980)
116. Best, F. R., Driscoll, M. J. (Eds.): Rep. MIT-EL-80-031 (1980)
117. Miyazaki, H.: Kobunshi Tokyo *30*, 193 (1981)
118. Baracco, L., Degetto, S., Marani, A., Croatto, U.: La Chimica e l'Industria *63*, 257 (1981)
119. Schwochau, K., Astheimer, L., Schenk, H.-J., Witte, E. G.: Chemiker Ztg. *107*, 177 (1983)
120. Spence, R.: Talanta *15*, 1307 (1968)
121. Wilson, J. D., Webster, R. K., Milner, G. W. C., Barnett, G. A., Smales, A. A.: Anal. Chim. Acta *23*, 505 (1960)
122. Miyake, Y., Sugimura, Y., Mayeda, M.: J. Oceanogr. Soc. Japan, *26*, 123 (1970)
123. Putral, A., Schwochau, K.: unpublished results
124. Ku, T. L., Knauss, K. G., Mathieu, G. G.: Deep Sea Research *24*, 1005 (1977)
125. Schwochau, K., Astheimer, L., Schenk, H.-J., Witte, E. G.: Z. Naturforsch. *37b*, 214 (1982)
126. De Bièvre, P., Linse, K. H., Schwochau, K.: To be published
127. Ogata, N., Inoue, N., Kakihana, H.: Nippon Genshiryoku Gakkaishi *13*, 560 (1971)
128. Graziani, R., Bombieri, G., Forsellini, E.: J. Chem. Soc. Dalton Trans. *1972*, 2059
129. Haigh, C. P.: Rep. R/M/N 787 (1976)
130. Schenk, H.-J., Astheimer, L., Witte, E. G., Schwochau, K.: Sep. Sci. Technol. *17*, 1293 (1982)
131. Davies, R. V., Kennedy, J., Peckett, J. W. A., Robinson, B. K., Streeton, R. J. W.: AERE-R-5024 (1965)
132. Ogata, N.: Genshiryoku Gakkaishi *11*, 82 (1969)
133. Astheimer, L., Schenk, H.-J., Witte, E. G., Schwochau, K.: unpublished results
134. Schwochau, K., Astheimer, L., Schenk, H. J.: 26. IUPAC-Congress Sept. 4–10, Tokyo 1977
135. Ogata, N.: Nippon Kaisui Gakkai-Shi *24*, 197 (1971)
136. Tani, H., Nakayama, H., Sakamoto, F.: Japan Kokai *75*, 134, 911 (1975)
137. Burba, P., Lieser, K. H.: Fresenius Z. Anal. Chem. *286*, 191 (1977)
138. Tabushi, I., Kobuke, Y., Ando, K., Kishimoto, M., Ohara, E.: J. Am. Chem. Soc. *102*, 5947 (1980)
139. Chen, Y. F.: Bull. Soc. Sea Water Science *36*, 24 (1981)
140. v. Falkai, B.: Synthesefasern, Verlag Chemie, Weinheim 1981, p. 236
141. Edgington, D. N., Gordon, S. A., Thommes, M. M., Almodovar, L. R.: Limnology and Oceanography, Baltimore *15*, 945 (1970)
142. Miyake, Y., Sugimura, Y., Mayeda, M.: J. Oceanogr. Soc. Japan *26*, 123 (1970)
143. Sakaguchi, T., Nakajima, A., Horikoshi, T.: Nippon Nogeikagaku Kaishi *53*, 211 (1979)
144. Astheimer, L., Schenk, H.-J., Witte, E. G., Schwochau, K.: Sep. Sci. Technol. *18*, 307 (1983)
145. Egawa, H., Harada, H.: Nippon Kagaku Kaishi *1979*, 958
146. Egawa, H., Harada, H., Shuto, T.: ibid. *1980*, 1767

147. Sugasaka, K., Katoh, S., Takai, N., Takahashi, H., Umezawa, Y.: Sep. Sci. Technol. *16*, 971 (1981)
148. Inoue, Y., Tsuji, M.: Bull. Chem. Soc. Japan *49*, 111 (1976)
149. Yamashita, H., Ozawa, Y., Nakajima, F., Murata, T.: ibid. *53*, 3050 (1980)
150. Bitte, J., Fremery, M., Bals, H. G.: Rep. MIT-EL-80-031 (1980)
151. Schwochau, K., Putral, A.: unpublished results
152. Research Commitee on Extraction of Uranium from Seawater. The Atomic Energy Soc. of Japan, Energy Developments in Japan, Vol. *3*, 67 (1980)
153. Dia-Prosim, Vitry, France, Technical Data Sheet 0100 A (1977)
154. Witte, E. G. et al.: to be published
155. Astheimer, L., Schenk, H.-J., Witte, E. G., Schwochau, K.: Sep. Sci. Technol. *18*, 307 (1983)
156. Kanno, M.: Report Nr. MIT-EL-80-031 (1980) p. 47 U
157. Nuclear Fuel, Febr. 4, 1980
158. Björk, B., Vallander, P.: Summary Report STU-Project 80-3955 (1981)
159. Bals, H. G.: Metal *33*, 401 (1979)
160. Bals, H. G.: Offenlegungsschrift DT 2550751 A1, 26. 05. 1977
161. Muzzarelli, R. A. A., Tubertini, O.: Talanta *16*, 1571 (1969)
162. Best, F. R.: Rep. MIT-EL-80-001 (1980)
163. Bitte, J., Graffunder, W.: Rep. UEB-3005-81-11-E (1981)
164. Kanno, M.: Sep. Sci. Technol. *16*, 999 (1981)
165. Bitte, J.: Uranerzbergbau, Bonn, Private Communication (May 1983)
166. Koske, P. H., Ohlrogge, K., Denzinger, H.: Meerestechnik *13*, 63 (1982)

Author Index Volumes 101–124

Contents of Vols. 50–100 see Vol. 100
Author and Subject Index Vols. 26–50 see Vol. 50

The volume numbers are printed in italics

Bei Fragen zur Produktsicherheit wenden Sie sich bitte an:
If you have any questions regarding product safety,
please contact:

Walter de Gruyter GmbH
Genthiner Straße 13
10785 Berlin
productsafety@degruyterbrill.com